the intelligence paradox

the intelligence paradox

why the intelligent choice isn't always the smart one

Satoshi Kanazawa

John Wiley & Sons, Inc.

Published by John Wiley & Sons, Inc., Hoboken, New Jersey
Published simultaneously in Canada

For general information about our other products and services, please contact our Customer Care Department within the United States at (800) 762-2974, outside the United States at (317) 572-3993 or fax (317) 572-4002.

Wiley also publishes its books in a variety of electronic formats and by print-on-demand. Some content that appears in standard print versions of this book may not be available in other formats. For more information about Wiley products, visit us at www.wiley.com.

Library of Congress Cataloging-in-Publication Data:

Kanazawa, Satoshi.

The intelligence paradox : why the intelligent choice isn't always the smart one / Satoshi Kanazawa.
p. cm.
Includes bibliographical references and index.
ISBN 978-0-470-58695-2 (hardback); ISBN 978-1-118-13764-2 (ebk); ISBN 978-1-118-13765-9 (ebk); ISBN 978-1-118-13766-6 (ebk)
1. Intellect. 2. Evolutionary psychology. I. Title.
BF431.K366 2012
153.9—dc23

2011042289

Printed in the United States of America

10 9 8 7 6 5 4 3 2 1

For Kaja Perina,
who makes all great things happen in my career

Contents

Acknowledgments

I first became interested in the problem of values when I was a student of Michael Hechter's at the University of Arizona. For my Ph.D. dissertation, I pursued an old theoretical problem that Michael had been interested in for many years, the Hobbesian problem of order (How is society possible among self-interested rational actors?). However, by the time I became his student in 1989, Michael was already beginning to move on to another project. He was trying to figure out where values and preferences of human actors came from. Rational choice theory can explain human behavior very well *if* it assumes what humans want and desire, but it cannot explain these wants and desires themselves. Michael, a leading rational choice theorist, wanted to advance the frontiers of rational choice theory by figuring out where human values and preferences came from, by providing a theoretical explanation for them. In technical language, he was trying to *endogenize* values in rational choice models.

I was not involved in Michael's values project at all, but was aware of the problem that he was struggling to solve. Even though Michael and I never discussed these theoretical issues, I was able to follow the contours and directions of his current thinking from the books and journal articles that he asked me to fetch from the library. On my long walks back from the main university library to the Social Sciences building, I would peruse the books and articles he wanted to read next. Sometimes I kept them for a couple of hours so that I could finish reading them before handing them to Michael.

Neither Michael nor I (nor anyone else) solved the problem of values, but the importance of the problem stayed with me. Years later, when I discovered evolutionary psychology by reading Robert Wright's *The Moral Animal*, I immediately realized that evolutionary psychology provided the answer that Michael had been searching for all these years. It could explain where human values and preferences came from. This book is the result of the realization.

My first and foremost intellectual debt is therefore to Michael Hechter, who unwittingly and unintentionally made me realize the importance of the problem of values and set me on this intellectual path. But I'm not sure how happy he is with the answer that I have found.

My intellectual life in London would be entirely barren but for a small group of like-minded scientists in and around London with whom I can share our mutual interest in evolutionary psychology and intelligence research. I thank Tomas Chamorro-Premuzic, Bruce G. Charlton, David de Meza, Adrian Furnham, Richard Lynn, Diane J. Reyniers, Jörn Rothe, Peter D. Sozou, and in particular Jay Belsky (who has since left London for greener intellectual pastures back in the United States) for years of stimulating discussions.

Since February 2008, I have had a blog at *Psychology Today* called *the Scientific Fundamentalist* (http://www.psychologytoday.com/blog/the-scientific-fundamentalist/). I've thrown out some of the ideas contained in this book initially on the blog, and they were subsequently developed, often with intellectual contributions from astute readers. I have a great team of editors at *Psychology Today* who have given tremendous support to me and my blog over the years. My warmest thanks go to Kaja Perina, Hara Estroff Marano, Carlin Flora, Lybi Ma, Wendy Paris,

PT COO Charles Frank, *PT* Director of Business Development Batya Lahav, and *PT* CEO Jo Colman. I consider them all to be part of my extended intellectual family.

Kaja is responsible for recruiting me as one of the inaugural *PT* bloggers. In December 2007, "blogging" was the last thing in the world I wanted to do. I had always thought that there could not possibly be any value in doing something that anyone can (and apparently does) do, and that "bloggers" were the most idiotic and self-obsessed group of people. (After four years, I have not changed my opinion of bloggers. The only people who are stupider than bloggers are people who leave anonymous comments on blogs.) With one transatlantic phone call, Kaja changed my mind and got me to agree to sign up as one of the four inaugural *PT* bloggers. As it turns out, to my great surprise, blogging was something I could do, and I was somewhat good at it. As usual, Kaja knew me much better than I knew myself.

In the last nine years, since Kaja became Editor-in-Chief, *Psychology Today* has done more to popularize and promote evolutionary psychology to the general nonacademic public than we evolutionary psychologists have done ourselves. So the scientific community of evolutionary psychologists owes a great intellectual debt of gratitude to Kaja and her superb editorial staff at *Psychology Today*.

A large number of friends and colleagues have made substantive contributions to this book, either as coauthors of the academic papers that constitute individual chapters in the book or generous colleagues who have taken the time to read and comment on drafts of such papers. I want to thank Jay Belsky, Tomas Chamorro-Premuzic, David de Meza, Ian J. Deary, Paula England, Aurelio José Figueredo, Barbara L. Finlay, Jeremy Freese, Linda S. Gottfredson, Josephine E. E. U. Hellberg, Christine Horne, Evelyn Korn, Norman P. Li, Patrick M. Markey, John D. Mayer, Andrew Oswald, Nando Pelusi, Kaja Perina, Qazi Rahman, Diane J. Reyniers, Todd K. Shackelford, Brent T.

Simpson, and Pierre L. van den Berghe. Gordon G. Gallup Jr. was kind enough to read the entire book manuscript and provide extremely constructive comments. Thanks to Kipling D. Williams for giving me permission to use a picture from Cyberball in Chapter 2 (and for inventing Cyberball in the first place).

None of my scientific research described in this book would have been possible without my formal affiliations with University College London and Birkbeck College University of London. I thank Adrian Furnham and David Shanks for offering me an honorary research fellowship in the Department of Psychology at UCL and Jay Belsky and Mike Oaksford for doing the same with the Department of Psychological Sciences at Birkbeck.

Finally, I would like to thank the pioneers in intelligence research, who have been unafraid to tell the truth in the face of political oppression and persecution. Many of them have paid tremendous personal and professional prices for their scientific integrity, but their courage has made it easier for others like me to continue their work and tell the truth. For their courageous and groundbreaking work, I thank Arthur R. Jensen, Richard Lynn, J. Philippe Rushton, Linda S. Gottfredson, Richard J. Herrnstein, Charles Murray, and Helmuth Nyborg. I am proud to count many of them among my personal friends and colleagues.

On 15 October 2001, in front of the building in Washington DC that houses the American Enterprise Institute, Charles Murray calmly told me that it would ultimately be my personal choice what kind of work I do, whether I succumb to the public pressure or pursue the truth no matter what the cost. For him personally, Charles continued, the life and career would not be worth living unless one told the truth. I was then too young, too inexperienced, too untenured, and too afraid to appreciate his wisdom. But I now know he was right. *E pur si muove.*

Introduction

This book challenges some of the common misconceptions that people have about the nature of intelligence: what it is, what it does, and what it is good for (if anything). People have a tendency to equate intelligence with character, and to believe that intelligence is the ultimate gauge of an individual's worth. They believe that people are not worthy human beings unless they are intelligent, at least in some way. Somehow, intelligence is regarded as the hallmark of human worth and character, the most important trait that any human can have.

Conversely, people claim that, because all humans are equally worthy, they must all hold the potential to be equally intelligent as well. They balk at scientific discoveries that suggest that some groups of individuals on average may not be as intelligent as others, as if that is tantamount to saying that such groups of people are somehow less worthy humans because of their lower intelligence. When they realize that some people do not score high on standard IQ tests, they automatically assume that IQ tests must therefore be biased against some groups of people. They take group differences in performance on the IQ tests as *prima facie* evidence of such bias.

At the same time, they also maintain that there are "multiple intelligences"[1] and fabricate other types of "intelligences" such as "bodily-kinetic intelligence" for people who are athletic, and "interpersonal intelligence" for extroverted people who have social skills. No longer does it suffice to say that some people are athletic and others are social; they must all possess different kinds of "intelligences." They have to ensure that everybody is intelligent in some way because everybody is an equally worthy human being.

In this book, I want to break this equation of intelligence with human worth, by pointing out that intelligence (and intelligent people) may not be what you think. While more intelligent people can do many things better and more efficiently than less intelligent people, there are many things that they cannot, and intelligent people tend to fail at the most important things in life from a purely biological perspective. The list of what intelligent people are *not* good at may surprise you. Intelligent people are only good at doing things that are relatively new in the course of human evolution. They are not necessarily good at doing things that our ancestors have always done, like finding and keeping a mate, being a parent, and making friends. Intelligent people tend not to be good at doing things that are most important in life.

There is no question that intelligence is a positive trait, but then so are beauty, height, and health. Yet we don't equate beauty, height, and health with human worth. (However, we do a little bit when it comes to beauty, by maintaining that people who are not physically attractive nonetheless have "inner beauty." "Inner beauty" is to physical attractiveness what "multiple intelligences" are to intelligence.) We don't necessarily think that beautiful, tall, or healthy people are better, more worthy humans than ugly, short, or unhealthy people. Nor do we claim that everyone is equally beautiful, equally tall, or equally healthy. But we seem to believe that more intelligent people are more worthy human beings. Or, conversely, because all humans ought to be equally worthy, they must all be equally intelligent. In this book, I want to convince you to stop having this assumption and break the equation of intelligence with human worth. Intelligence is just another quantitative trait of an individual like height or weight.

What Do People Want?

This book is also about what people want—their preferences and values. What do people want? Why do people want what they want? Where do individual preferences and values come from? These are some of the questions that I address in this book.

These questions are known as the "problem of values," and it is one of the central questions in social and behavioral sciences. Mathematically elegant models of human behavior, especially in microeconomics, can explain human behavior very well, *if* we know what people want in the first place. These microeconomic models essentially say, "Human actors do the best they can within their constraints and circumstances to achieve what they want to achieve," and their mathematical models are about how these constraints and circumstances affect their choices. However, these models cannot tell us what human actors actually do and what actual choices they make, unless we know "what they want to achieve," namely, their goals, preferences, and values.

Without the knowledge of individual preferences and values, *any* choice can be used to support the microeconomic model. Let's say two individuals, A and B, are facing identical constraints and circumstances; they both find themselves at a local grocery store, with limited choices of items to buy and only $10 to spend. If, under these conditions, Individual A purchases apples and Individual B purchases oranges, both choices can be used to support the rational actor model, by positing, *ex ante,* that Individual A has a preference for apples while Individual B has a preference for oranges. If, under the same conditions, Individual C chooses not to buy anything at all, that, too, supports the theory because Individual C has a preference for savings.

In order to provide a full explanation of human behavior, we must *endogenize* individual preferences and values, by making them part of the theoretical explanations. We must first explain individual preferences and values *before* we can explain the behavioral choices. We must be able to explain *why* Individual A wants apples, not oranges, and why Individual B wants oranges, and not apples, and why Individual C wants neither.

However, economists themselves have given up on the attempt to explain preferences and values. In 1977, two University of Chicago economists (and future Nobel prizewinners), George J. Stigler and Gary S. Becker, published a very influential paper in the *American Economic Review* entitled "De Gustibus Non Est Disputandum."[2] The Latin title means "There's no accounting for tastes." In this paper, Stigler and Becker argued that economics should not be concerned with explaining preferences and values, but instead with how constraints affect choice. In other words, they wanted to leave preferences and values *exogenous* to the economic models, rather than *endogenous* to them. (*Endogenous* means within the model of explanation whereas *exogenous* means outside of such models. Endogenous factors are explained within the model whereas exogenous factors are left unexplained.)

Economists since then have faithfully followed Stigler and Becker's dictum and left preferences and values unexplained. They focus on money—how individuals seek to maximize income and wealth—because money is fungible[3]; people can use money to buy virtually anything they want (except, of course, for love). Economists reason that they don't need to know people's preferences and values because they would all want to maximize monetary incomes *regardless* of their idiosyncratic preferences and values. Whatever they want, they all want more money because more money will allow them to have more of whatever they want. With more money, you can buy more apples, more oranges, or increase the level of your savings. So, in the above example, Individuals A, B, and C—in fact, *all* individuals—would want to maximize their monetary income because it would allow each to purchase whatever they want. The focus on the fungible resource of money obviates the need to know what individuals *really* want. This is where microeconomics stands today.

I want to endogenize preferences and values to models of human behavior, by introducing evolutionary psychology. I want to explain what people want and why they want it, and I believe evolutionary psychology is key to such an explanation. So this book is also about explaining the origin of individual preferences and values.

Given the main focus on intelligence in this book, I will emphasize the role that intelligence plays in influencing people's preferences and values. How does intelligence influence people's preferences and values? What do more intelligent people want? What do less intelligent people want? And why are there differences?

So this book is at the intersection of two scientific fields: evolutionary psychology and intelligence research. In Chapter 1, I briefly introduce you to the field of evolutionary psychology. Those of you who have read my previous book *Why Beautiful People Have More Daughters*[4] can probably skip Chapter 1 without missing much.

In Chapter 2, I discuss one important implication of evolutionary psychology for the functions and constraints on the human brain. I introduce an important idea called *the Savanna Principle* about what the human brain can easily do or has difficulty doing. Readers of my earlier book may recall reading about the Savanna Principle in it.

In Chapter 3, I discuss the concept of intelligence and introduce you to the field of intelligence research. I will explain what intelligence is, what it is not, and what it does (and doesn't do). There are so many misconceptions about intelligence, and I will try to dispel many of them in this chapter.

In Chapter 4, I introduce you to the central idea in this book, which I call "the Intelligence Paradox." It is about how more intelligent people may differ from less intelligent people in their preferences and values, often in quite counterintuitive ways. Readers of my *Scientific Fundamentalist* blog at *Psychology Today* will see that "the Intelligence Paradox" captures the same essential idea as what I call "the Hypothesis" in my blog, but applied specifically to individual preferences and values.

The rest of the book (Chapters 5–12) details various applications and manifestations of the Intelligence Paradox. Chapter 5 is about political attitudes on the liberal-conservative continuum, and explains why liberals in the United States are on average more intelligent than conservatives. I also explain why liberals are stupider than conservatives. (No, that's not a typo. Confused? Read on!)

Chapter 6 is about religiosity and belief in God. I will explain where religion came from and why atheists are more intelligent than people who believe in God. It's not because religion is false or there is no God. It's because religion is deeply human. We are designed by evolution to believe in God, and that is why more intelligent people are atheists.

Chapter 7 is about preferences about sexual exclusivity. I will explain why more intelligent men prefer being sexually exclusive more than less intelligent men, but more intelligent women do not value sexual exclusivity more than less intelligent women. People think that evolutionary psychology is all about sex differences between men and women, but, in this book, this is the only area in life where the Intelligence Paradox makes different predictions for men and women. The chapter also explains why, despite their greater preference for sexual exclusivity, more intelligent men are nonetheless more likely to cheat on their wives and girlfriends than less intelligent men. Once again, that's not a typo. More intelligent men are simultaneously more likely to value being sexually faithful and more likely to cheat.

Chapter 8 is about circadian rhythms, whether you are a morning person or a night person. I will explain why night owls are more intelligent than morning larks. Believe it or not, more intelligent people go to bed and wake up later than less intelligent people, and it's because it's unnatural for humans to stay up late at night.

Chapter 9 is about sexual orientation, where I will explain why homosexuals are on average more intelligent than heterosexuals. Sexual orientation, at least for men, is largely genetically and hormonally determined before birth. But, within their genetic constraints, more intelligent individuals choose to engage in homosexual behavior more than less intelligent individuals. Genetic influence and individual choice are not necessarily incompatible.

Chapter 10 deals with people's musical preferences. It is generally assumed that more intelligent people prefer certain types of music, like classical music. But why? Why should intelligence affect people's musical preferences? And what other types of music do intelligent people prefer? The answer may surprise you. It turns out that intelligent people also like elevator music.

Chapter 11 is about substance use: alcohol, tobacco, and drugs. Consistent with the Intelligence Paradox, there appears to be some evidence that more intelligent people drink more alcohol, smoke more cigarettes, and use more illegal drugs than less intelligent people. Not only are more intelligent people more likely to consume alcohol more frequently and in greater quantities, they are also more likely to binge drink and get drunk. Yes, more intelligent people are more likely to do stupid things. I will explain why in this chapter.

Chapter 12 is about how intelligent people fail at the most important task in life, which is reproductive success. This is where I make the point that intelligent people are the ultimate losers in life. (Yes, that's an evolutionary pun.) They are really not that great when it comes to important things that truly matter in life. There is some evidence that more intelligent people—at least more intelligent women—have fewer children and are altogether less likely to become parents. I will also discuss what this finding likely means for the average intelligence of the population of the future generations in western industrial societies.

In Chapter 13, I explore some of the other possible implications of the Intelligence Paradox, and wonder what other values may be affected by intelligence. For example, it has been known that criminals on average have lower intelligence than law-abiding citizens. But why? And why are vegetarians more intelligent than meat eaters?

In this book, I continue the system of references that I borrowed in a modified form from Charles Murray[5] and that I began using in *Why Beautiful People Have More Daughters*. For endnotes which simply give citation information, I use the standard endnote reference numbers, like this.[1] References which give greater information not contained in the text have reference numbers with brackets, for example.[[2]]

A Brief Word on the Data

Most of the empirical analyses that are summarized in this book use three different data sets: General Social Surveys (GSS) in the US, National Longitudinal Study of Adolescent Health (Add Health) in the US, and the National Child Development Study (NCDS) in the UK. They are all large, nationally representative, high-quality samples. Two of them are prospectively longitudinal data which have followed a cohort of individuals for many years, in one case, since birth for more than half a century. Here I will briefly describe the data sets and how each survey measures intelligence, which is the central concept in this book. Interested readers who want to know more about the actual empirical analyses can consult the academic papers cited in the chapters.

General Social Surveys (GSS)

The GSS is widely considered to be the best source of data in the world on a wide range of social attitudes and social trends. Its success has spawned counterparts in other nations (such as Canada, the UK, the Netherlands, and Germany), which now conduct their own periodic surveys of social attitudes which are modeled after the GSS in the US.

The National Opinion Research Center at the University of Chicago has administered the GSS, either annually or, more recently, biennially, since 1972. Personal interviews are conducted with a nationally representative sample of non-institutionalized adults over the age of 18 in the US. The sample size is about 1,500 individuals for each annual survey, and about 3,000 individuals for each biennial one. The exact questions asked in the survey vary by the year. The 1972–2008 cumulative data file contains 53,043 individuals and 5,364 variables (although not all variables are measured of all individuals).[6]

The GSS measures the respondent's verbal intelligence with a 10-item synonyms test. The survey asks the respondent to select a synonym for a word out of five candidates. Half of the respondents in each GSS sample answer 10 of these questions. This is a measure of verbal intelligence, not strictly general intelligence, which is the focus of this book. However, verbal intelligence is known to be very highly correlated with general intelligence.[7] In fact, as I explain in Chapter 3, all measures of intelligence are highly correlated with each other.

National Longitudinal Study of Adolescent Health (Add Health)

The National Longitudinal Study of Adolescent Health (Add Health) is a prospectively longitudinal study of a nationally representative sample of more than 20,000 adolescents in junior high and high school (Grades 7–12 in the 1994–1995 school year) in the United States. The Add Health cohort has been followed into young adulthood with four in-home interviews (in 1994–1995, 1996, 2001–2002, and 2008).[8]

Add Health measures the respondent's verbal intelligence with an abbreviated version of the Peabody Picture Vocabulary Test (PPVT). The PPVT asks the respondent to identify a picture that corresponds to a word out of four candidate pictures. While properly a measure of verbal intelligence, the PPVT has been shown to be a good measure of general intelligence as well.[9] For all analyses of Add Health in this book, the intelligence is measured during adolescence (at Wave I) whereas all the outcome measures are taken in early adulthood in Wave III.

National Child Development Study (NCDS)

The National Child Development Study (NCDS) is one of the world's oldest prospectively longitudinal studies. It has followed a population of respondents in the United Kingdom since birth for more than half a century. The study includes *all* babies (more than 17,000) born in Great Britain (England, Wales, and Scotland) during one week (03–09 March 1958). The respondents have subsequently been reinterviewed eight times (at ages 7, 11, 16, 23, 33, 42, 47, and 51). In each survey, personal interviews and questionnaires are administered to the respondents, to their mothers, teachers, and doctors during childhood, and to their partners and children in adulthood.

The NCDS has one of the best measures of general intelligence of all large-scale, population-based surveys. The NCDS respondents take multiple intelligence tests at ages 7, 11, and 16. At age 7, the respondents take four cognitive tests (Copying Designs Test, Draw-a-Man Test, Southgate Group Reading Test, and Problem Arithmetic Test). At age 11, they take five cognitive tests (Verbal General Ability Test, Nonverbal General Ability Test, Reading Comprehension Test, Mathematical Test, and Copying Designs Test). At age 16, they take two cognitive tests (Reading Comprehension Test, and Mathematical Comprehension Test). From these 11 cognitive tests, a measure of childhood general intelligence is calculated by a statistical technique called factor analysis. The technique measures the underlying general intelligence that explains the individual's performance on all of these varied cognitive tests at three different ages. It also eliminates all random measurement errors which are inherent in any measurement of human traits.

Chapter 1

What Is Evolutionary Psychology?[1]

Evolutionary psychology, at the most fundamental level, is the study of human nature. Human nature consists of what evolutionary psychologists call *evolved psychological mechanisms* or *psychological adaptations* (which are roughly synonymous with each other). Evolved psychological mechanisms provide solutions to *adaptive problems* (problems of survival and reproduction). Through a long process of natural and sexual selection, evolution has equipped humans with the ability to solve important problems, by allowing those who could solve the problems to live longer and reproduce more successfully and by eliminating those who couldn't. Those who had these innate solutions in their brain enjoyed distinct advantages over those who didn't, and lived longer and produced more children who survived. And their children inherited their parents' genetic tendency to solve these problems, and, in turn, lived longer and had more children themselves.

Over time there were more and more people who had these solutions in their brains and fewer and fewer people who didn't, until these innate solutions to adaptive problems became universal, characterizing all normally developing members of the human species. Human nature is therefore universal or *species-typical* (typical or characteristic of all members of a species). Some evolved psychological mechanisms are specific to only men or only women; others are shared by both men and women.

The important point to remember is that the psychological adaptations produce the correct solutions to the adaptive problems *only in the context of the ancestral environment.* Evolved psychological mechanisms are designed for and adapted to the conditions of the ancestral environment, not necessarily to those of the current environment. Evolution cannot anticipate or foresee the future, so its products—evolved psychological mechanisms—are not necessarily adapted to the conditions that emerged after they were designed. To the extent that our current environment is radically different from the ancestral environment, where our ancestors lived on the African savanna as hunter-gatherers in a small band of about 150 related individuals,[2] then the execution of the evolved psychological mechanisms does not necessarily produce the correct solutions to the adaptive problems at hand. In fact, as you will see below, it often produces the wrong solutions.

Our ancestors were, and had been for more than a million years, hunter-gatherers, first in Africa, then elsewhere on earth. Their hunter-gatherer lifestyle came to an (evolutionarily speaking) abrupt end around 10,000 years ago, when agriculture was invented. The invention of agriculture at around 8,000 BC is probably the single most important event in human history. Agriculture necessitated sedentary life; our ancestors, for the first time, ceased to be nomadic and stayed put in one place. That led to permanent settlements, villages, towns, cities, houses, roads, horse carriages, bridges, buildings, governments, democracy, automobiles, airplanes, computers, and iPods. The iPods would not have been possible without agriculture and everything else it led to.

Four Core Principles of Evolutionary Psychology

Evolutionary psychology, in its intellectual origin, is the application of evolutionary biology to human cognition and behavior. Ever since Darwin, evolutionary biologists and zoologists had known that principles of evolutionary biology applied to all species in nature, *except* for humans. In 1992, a group of psychologists and anthropologists, following the courageous lead of E. O. Wilson,[3] simply asked "Why not?"[4] Why are humans exceptions to the rule of nature? Why not apply the same principles of evolutionary biology to humans as well? And thus evolutionary psychology was born, merely 20 years ago. It's a very new science. But it has made tremendous progress in its very short history.

As an application of evolutionary biology to human cognition and behavior, evolutionary psychology is based on four core principles.

1. People Are Animals[5]

The first and most fundamental principle of evolutionary psychology is that there is nothing special about humans. This realization, that humans are not exceptions to nature but part of it, initially led the original evolutionary psychologists to apply the laws of evolution by natural and sexual selection to humans. It turns out that humans are not exceptions to nature at all, but just another animal species.

Scientists once believed that humans possessed many traits that were strictly unique to humans and that no other species had, such as culture, language, tool use, consciousness, morality, sympathy, compassion, romantic love, homosexuality, murder, and rape. This turns out to be false. Recent scientific research has shown that there is at least one other species that shares any trait that humans have.[6] To the best of my knowledge, there are no traits that only humans have.

This, however, does *not* mean that humans are not unique. To quote the great sociobiologist Pierre L. van den Berghe, "Certainly we are unique, but we are not unique in being unique. Every species is unique and *evolved* its uniqueness in adaptation to its environment."[7] The fact that humans are unique means that no other species have the exact *constellation* of traits and characteristics that humans have. If chimpanzees were *exactly* the same as humans in every possible way, then they would not be a separate species from humans; they would be humans. Humans are a separate species because no other species is *exactly* like humans.

But this is true of every species in nature: dogs, cats, giraffes, cockroaches. No other species is exactly like cockroaches. Humans as a species are just as unique and special as cockroaches, no more, no less. Every species in nature is equally unique.

The unavoidable conclusion from evolutionary biology is that there is nothing special about humans as a species, and we are just another ape species in nature. As such, all laws of biology hold for humans as they do for all other species. And this includes the law of evolution by natural and sexual selection, which states that the ultimate goal of all living organisms is reproductive success. All living organisms in nature are designed by evolution to reproduce and make as many copies of their genes as possible.

2. There Is Nothing Special about the Human Brain

For evolutionary psychologists, the brain is just another body part, like the hand or the pancreas. Just as millions of years of evolution have gradually shaped the hand or the pancreas to perform certain functions, so has evolution shaped the human brain to perform its function, which is to solve adaptive problems to help humans survive and reproduce successfully. Evolutionary psychologists apply the same laws of evolution to the human brain as they do to any other part of the human body.

Social scientists tend to believe that evolution stops at the neck.[8] They believe that, while evolution has shaped the structure and function of every other human body part, the human brain has been immune to evolutionary history. In sharp contrast, evolutionary psychologists contend that the human brain is not an exception to the influences of evolutionary forces on the human body. Evolution does *not* stop at the neck; it goes all the way up.

3. Human Nature Is Innate

Just as dogs are born with innate dog nature, and cats are born with innate cat nature, humans are born with innate human nature. This follows from Principle 1 above. What is true of dogs and cats must also be true of humans. Socialization and learning are very important for humans, but humans are born with the innate capacity for cultural learning. Pierre van den Berghe continues the quote above by saying, "Culture is the uniquely human way of adapting, but culture, too, evolved biologically." Culture and learning are part of the evolutionary design for humans. Socialization merely reiterates and reinforces what is already in our brain (like the sense of right and wrong, which we share with other species[9]).

This principle of evolutionary psychology is in clear contrast to the blank slate ("tabula rasa") assumption held by most social scientists.[10] They contend that, because evolution stops at the neck, humans are born with a mind like a blank slate, on which cultural socialization must and can write anything whatsoever. Evolutionary psychologists strongly reject the tabula rasa assumption of the social sciences. In the memorable words of William D. Hamilton, who is universally regarded as the greatest Darwinian since Darwin, "The *tabula* of human nature was never *rasa* and it is now being read."[11] Evolutionary psychology is devoted to reading the tabula of human nature.

4. Human Behavior Is the Product of Both Innate Human Nature and the Environment

There are a few genetic diseases, such as Huntington's disease, that are 100% determined by genes. If someone carries the affected gene, they will develop the disease no matter what their experiences or environment.[12] An individual's eye color or blood type is also 100% determined by genes. So these (and a very few other) traits are entirely genetically determined. Otherwise, there are no human traits that are 100% determined by genes. Nor are there any serious scientists who think there are complex human behaviors that are entirely determined by genes. Contrary to what the critics of evolutionary psychology often claim, there are no genetic determinists in science.

Genes for most traits seldom express themselves in a vacuum. Their expressions—how the genes translate into behavior—often depend on and are guided by the environment. The same genes can express themselves differently depending on the context. In this sense, both innate human nature, which the genes program, and the environment in which humans grow up and live, are equally important determinants of behavior.

Many social scientists believe that human behavior is 100% determined by the environment, and genes and biology have absolutely no role to play in it.[13] In sharp contrast, evolutionary psychologists do not believe that human behavior is 100% determined by either genes or environment alone. However, evolutionary psychologists tend to emphasize the biological and genetic factors in their research, because they are fighting the supremacy of *environmentalism* (the belief that the environment determines human behavior 100%) both in the social sciences and among the general public. Nobody is surprised to learn that the environment influences behavior; that is not news. But people are often surprised by the extent to which genes influence behavior. That is news.

Two Logical Fallacies That We Must Avoid

In any discussion of evolutionary psychology, or human sciences in general, it is very important to avoid two logical fallacies. They are called the naturalistic fallacy and the moralistic fallacy.

The *naturalistic fallacy*, which was coined by the English philosopher George Edward Moore[14] in the early 20th century, though first identified much earlier by the Scottish philosopher David Hume,[15] is the leap from *is* to *ought*—that is, the tendency to believe that what is natural is good; that what is, ought to be. For example, one might commit the error of the naturalistic fallacy and say, "Because different groups of people *are* genetically different and endowed with different innate abilities and talents, they *ought* to be treated differently."

The *moralistic fallacy*, coined by the Harvard microbiologist Bernard Davis[16] in the 1970s, is the opposite of the naturalistic fallacy. It refers to the leap from *ought* to *is*, the claim that the way things ought to be is the way they are. This is the tendency to believe that what is good is natural; that what ought to be, is. For example, one might commit the error of the moralistic fallacy and say, "Because everybody *ought* to be treated equally, there *are* no innate genetic differences between groups of people." The science writer extraordinaire Matt Ridley calls it the *reverse naturalistic fallacy*.[17]

Both are logical fallacies, and they get in the way of progress in science in general, and in evolutionary psychology in particular.

However, as Ridley astutely points out, political conservatives are more likely to commit the naturalistic fallacy ("Nature designed men to be competitive and women to be nurturing, so women ought to stay home to take care of the children and leave business and politics to men."), while political liberals are equally likely to commit the moralistic fallacy ("The Western liberal democratic principles hold that men and women ought to be treated equally under the law, and therefore men and women are biologically identical and any study that demonstrates otherwise is *a priori* false."). The evolutionary psychologist Robert O. Kurzban concisely captures the common attitude among political liberals when he quips, "It's only 'good science' if the message is politically correct."[18]

Since academics, and social scientists in particular, are overwhelmingly left-wing liberals, the moralistic fallacy has been a much greater problem in academic discussions of evolutionary psychology than the naturalistic fallacy. Most academics are above committing the naturalistic fallacy, but they are not above committing the moralistic fallacy. The social scientists' stubborn refusal to accept sex and race differences in behavior, temperament, and cognitive abilities, and their tendency to be blind to the empirical reality of stereotypes, reflect their moralistic fallacy driven by their liberal political convictions.

The left-wing denial of certain inconvenient empirical truths culminates in the wholesale postmodern denial of scientific objectivity and the concept (and possibility) of scientific truth. Conservatives too deny *some* empirical truths, like evolution, but they do not deny that there is such a thing as a scientific truth. But, once again, we do not have to worry about conservatives in academia, because there are very few of them (and you will find out why in Chapter 5). There are virtually no creationists who deny evolution among the faculties of American universities, but there are many, many postmodernists who deny scientific objectivity.

It is actually very easy to avoid both fallacies—both leaps of logic—by simply *never* talking about what ought to be at all and only talking about what is. It is not possible to commit either the naturalistic or the moralistic fallacy if scientists never talk about *ought*. Scientists, by which I mean *basic* scientists, not *applied* scientists like engineers and physicians, do not draw moral conclusions and implications from the empirical observations they make, and they are not guided in their observations by moral and political principles. Real scientists—basic scientists—only care about what is, and do not at all care about what ought to be. In this book, I will only talk about what is, and will never talk about what ought to be.

What Does "Natural" Mean?

It is always important in any discussion of science to avoid the naturalistic and moralistic fallacies, but it is particularly important to remember not to be tempted by them when you read this book. From a purely biological perspective, *natural* only means "that for which the organism is evolutionarily designed" and *unnatural* only means "that for which the organism is not evolutionarily designed." The only organisms that I will talk about in this book are humans. From a purely scientific perspective, murder[19] and rape[20] are completely natural for humans, and getting a Ph.D. in evolutionary psychology is completely unnatural (which is partly why intelligent people do it!). *Natural* decidedly does *not* mean *good*, *valuable*, or *desirable*, and *unnatural* decidedly does *not* mean their opposites. One of the consistent themes in this book is that intelligent people often do unnatural things.

There are only two legitimate criteria by which to evaluate scientific ideas and theories: logic and evidence. Accordingly, you may justifiably criticize evolutionary psychological theories (or any other theories in science) if they are logically inconsistent within themselves or if there is credible scientific evidence against them. As a scientist, *as the Scientific Fundamentalist*,[21] I take all such criticisms seriously. However, you may not criticize scientific theories, whether mine or otherwise, simply because their implications are immoral, ugly, contrary to our ideals, or offensive to some or all. I can tell you right now that the implications of many of the scientific ideas and theories discussed in this book, whether mine or otherwise, are indeed immoral, ugly, contrary to our ideals, or offensive to some group of people. They are very offensive to me. But it doesn't matter.

The truth is the *only* guiding principle in science, and it is the most important thing for all scientists. In fact, it is the *only* important thing; nothing else matters in science besides the truth. However, I also believe that any solution to a social problem must start with the correct assessment of the problem itself and its possible causes. We can never devise a correct solution to a problem if we don't know what its ultimate causes are. So the true observations are important foundations of both basic science and social policy, if you do care about solving social problems, which of course I don't.

If what I say is wrong (because it is illogical or lacks credible scientific evidence), if it is not true, then it is my problem. It is my job as a scientist then to construct better theories or collect more evidence. In contrast, if what I say offends you, it is your problem. My credo as a scientist, which undergirds my scientific fundamentalism, is "If the truth offends people, it is our job as scientists to offend them."[22] As a scientist, as the Scientific Fundamentalist, I don't care if people live or die. I just want to know why.

Chapter 2

The Nature and Limitations of the Human Brain

In this chapter I will focus on the human brain as an evolved organ and talk about how evolution has shaped and designed the human brain to have certain limitations and constraints. I will introduce you to a very fundamental observation in evolutionary psychology, which I call "the Savanna Principle."[1]

The Savanna Principle

Evolutionary adaptations, whether they are physical or psychological, are designed for and adapted to the conditions of the ancestral environment, during the period of their evolution, not necessarily to the conditions of the current environment.[2] Evolution cannot anticipate or foresee the future; it can only respond to conditions in the past. So it is impossible for it to design adaptations that will suit conditions that have not yet existed. This is easiest to see in the case of physical adaptations, such as the vision and color-recognition system.

What color is a banana? A banana is yellow in the sunlight and in the moonlight. It is yellow on a sunny day, on a cloudy day, on a rainy day. It is yellow at dawn and at dusk. The color of a banana appears constantly yellow to the human eye under all these conditions, despite the fact that the actual wavelengths of the light reflected by the surface of the banana under these varied conditions are different. Objectively, they are not the same color all the time. However, the human eye and color-recognition system can compensate for these varied conditions because they all occurred during the course of the evolution of the human vision system, and the visual cortex can perceive the objectively varied colors as constantly yellow.[3]

So a banana looks yellow under all conditions, *except in a parking lot at night.* Under the sodium vapor lights commonly used to illuminate parking lots, a banana does not appear natural yellow. This is because the sodium vapor lights did not exist in the ancestral environment, during the course of the evolution of the human vision system, and the visual cortex is therefore incapable of compensating for them. Evolution, which designed the human eye, only "knew" about the sun, the moon, and possibly open fire as the only sources of illumination. It could not have anticipated the sodium vapor lights or any other type of artificial illumination, like fluorescent lamps, which is why things look unnatural under fluorescent light.

Fans of the 1989 James Cameron movie *The Abyss* may recall a scene toward the end of the movie, where it is impossible for Ed Harris's character (a deep-sea diver) to distinguish colors under artificial lighting in the otherwise total darkness of the deep oceanic basin. Regular viewers of the TV show *Forensic Files* (formerly known as *Medical Detectives*) and other real-life crime documentaries may further recall that eyewitnesses often misidentify the colors of cars on freeways, leading the police either to rule in or rule out potential suspects incorrectly. This happens because highways and freeways are often lit with sodium vapor lights and other evolutionarily novel sources of illumination, which distort colors to the human eye.

The same principle that holds for physical adaptations like the color recognition system also holds for psychological adaptations. Pioneers of evolutionary psychology[4] all explicitly recognized that the evolved psychological mechanisms are designed for and adapted to the conditions of the ancestral environment, not necessarily to those of the current environment. In 2004, I systematized these observations into what I call the *Savanna Principle.*[5]

> *Savanna Principle: The human brain has difficulty comprehending and dealing with entities and situations that did not exist in the ancestral environment.*

Other evolutionary psychologists call the same observation *the evolutionary legacy hypothesis*[6] or *the mismatch hypothesis.*[7] The names may vary, but the observation remains the same. We are stuck with the stone-age brain which assumes that we are still hunter-gatherers on the African savanna, and responds to the environment as if it were the African savanna.[8] There are many manifestations of the Savanna Principle in our daily life.

TV Friends

Here is one illustration of the Savanna Principle in action. In 2002, I discovered that individuals who watched certain types of TV shows were more satisfied with their friendships, just as they were if they had more friends or socialized with them more frequently.[9] And this finding was later replicated by others.[10] From the perspective of the Savanna Principle, this may be because realistic images of other humans, such as those found in TV, movies, videos, photographs, and DVDs, did not exist in the ancestral environment, where all realistic images of other humans *were* other humans. As a result, the human brain may have implicit difficulty distinguishing their "TV friends" (the characters they repeatedly see on TV shows) and their real friends. So by being repeatedly exposed to their "TV friends," that is, by watching TV shows with their familiar characters, they feel like they are actually with their friends, and their satisfaction with friendships increases.

The Savanna Principle suggests that, because TV and other realistic electronic depictions of other human beings did not exist in the ancestral environment, our brain cannot really comprehend TV. A lot of people get angry when I say this, and they vehemently deny that their brain fails to distinguish between "TV friends" and their real friends. They adamantly insist that they *do* know how TV works.

Well, you do and you don't. At the conscious level, you do know how TV programs are produced. You do know that the people you see on TV are actors, who are hired and paid millions of dollars to play certain roles written by screenwriters in their scripts. You consciously know that TV shows and movies aren't real.

But your brain doesn't. If it does, why did you cry when Julia Roberts's character died at the end of *Steel Magnolias*? Don't you know that she's just an actor who was paid a lot of money (reportedly, $90,000 in 1989[11]) to play the role of a dying woman? Don't you know that she's not really dead? Why did you get scared when Freddie Kruger slashed many teenagers in *A Nightmare on Elm Street*? Don't you know that the actor who played Freddie Kruger (Robert Englund) is really a nice person and has never killed anyone? Don't you know that none of the teenagers who were murdered in the movie actually died, nor were they in any real danger because they were surrounded by dozens of crew at every moment? Your brain doesn't truly comprehend any of these things, and that's why you are able to enjoy watching movies and TV shows. If your brain *truly did* comprehend movies and TV shows, you would never be able to enjoy them.

This is what I mean by *comprehension* in the Savanna Principle (and later in the Savanna-IQ Interaction Hypothesis, discussed in Chapter 4). Comprehension means true, logical, and scientifically and empirically accurate understanding of how something works. Your brain (as opposed to you) truly comprehends something when your reaction or behavior in response to it is consistent with a scientifically and empirically accurate understanding of how it works.

So true comprehension of a TV show is that a large number of professional actors are paid to enact certain scripted roles but the characters they play on the show do not really exist in real life. Untrue comprehension of it includes, among others, that the stories portrayed on the show are real and that you know the characters on the show personally and they know you personally. The studies discussed above suggest that your brain (as opposed to you) does not always have true comprehension of TV shows because your reaction to them—increased satisfaction with your friendships—suggests otherwise.

Pornography

Pornography—in particular, the vast sex differences in its consumption and reactions to it—is another illustration of the Savanna Principle at work.

An overwhelming majority of consumers of pornography worldwide are men.[12] Given their greater desire for sexual variety, it is understandable why men would consume more pornography and seek out sexual encounters with numerous women in pornographic photographs and videos, just as they do when they contract prostitutes in search of greater sexual variety.

Such desire for sexual variety on the part of men is evolutionarily adaptive. A man who has sex with 1,000 women in a year can potentially produce 1,000 children (or more if there are multiple births); more realistically, he can expect to father 30 children (given that the probability of conception per coital act is .03).[13] In sharp contrast, a woman who has sex with 1,000 men in a year can still only expect to have one child (barring a multiple birth) in the same time period, which she can achieve by having regular sex with only one man. So there is very little reproductive benefit for women in seeking a large number of sex partners, as there is for men.

However, unlike consorting with prostitutes, watching pornography does not lead to actual sexual intercourse. So why do men like to consume pornography?

The Savanna Principle suggests that a man's brain does not really know that he cannot copulate with the women he sees in pornographic photographs and videos. When men see images of naked and sexually receptive women, their brains cannot truly comprehend that they are artificial images of women whom they will likely never meet, much less have sex with, because no such images existed in the ancestral environment. Every single naked and sexually receptive woman that our male ancestors saw throughout human evolutionary history was a potential sex partner.

As a result, their brains think that they might have actual sexual encounters with these women. Why else would men have an erection when they view pornographic photographs and videos, when the only biological function of an erection is to allow men to have intercourse with women? If men's brains truly comprehended that they would likely never have sex with the naked women in pornography, they would not get an erection when they watch it.

The same principle holds in strip clubs and peep shows, even though these involve real *live* women, not their photographic and electronic images. In the ancestral environment, there were no women who were *paid* to dance around naked in front of men, and *pretended* to be sexually aroused and interested in them, but would never actually have sex with them. So men's brains cannot truly comprehend strippers and lap dancers. That is why they get an erection at strip clubs and peep shows, when they consciously know that they would not actually have sex with the naked women dancing in front of them.

The failure of men's brains to comprehend images of naked women nevertheless has some real consequences. In one experiment, men who viewed *Playboy* centerfolds subsequently found their own girlfriends physically less attractive and expressed less love for them.[14] Why would men love their girlfriends less after viewing pictures of naked women in *Playboy* unless their brains implicitly assume that they could potentially date the *Playboy* centerfolds instead of their current girlfriends, most of whom pale in comparison to them?

But it's not only men's brains that fail. The Savanna Principle applies equally to women as it does to men; women's brains have the same evolutionary constraints and limitations as those of men. This is why women do not consume pornography nearly as much as men do, even though women enjoy having sexual fantasies as much as men do.[15] Women do not seek sexual variety because their reproductive success does not increase by having sex with a large number of partners. In fact, given the limited number of children they can have in their lifetimes, the potential cost of having sex with the wrong partner is far greater for women than it is for men. This is why women are far more cautious about having sex with someone they do not know well[16] and tend to require a much longer period of acquaintance before agreeing to have sex than men do.[17]

So it makes perfect evolutionary sense for women to avoid casual sex with anonymous strangers, and their brains cannot really tell that there is no chance that they might copulate with, or, worse yet, be impregnated by, a large number of the naked and sexually aroused men they see in pornography. Women's brains cannot fully comprehend that they will not get pregnant by watching pornography, just as men's brains do not know that they cannot copulate with women in pornography. Women avoid pornography for the same reason that men consume it; in both cases, their brains cannot really distinguish between real sex partners and the imaginary ones, because the latter did not exist in the ancestral environment.

The human brain (of both men and women), incidentally, also implicitly assumes that all sex potentially leads to reproduction. That is why we still experience sex with contraception as pleasurable, and we are motivated to pursue it. The human brain, adapted to and designed for the ancestral environment, cannot truly comprehend modern contraceptives that did not exist in the ancestral environment. If it did, we would not find sex with contraception physically pleasurable. The only contraception that existed in the ancestral environment was abstinence, and, consequently, this is the only form of contraception that the human brain can truly comprehend. This why men and women do not experience abstinence (not having sex) as pleasurable, but still experience sex when the woman is on the pill as pleasurable, even though both have the same reproductive consequences.

Cooperation in One-shot Prisoner's Dilemma Games

In a Prisoner's Dilemma game, two players make simultaneous decisions without knowing what the other player decides. Each player can decide to "cooperate" with the other player, or to "defect" on the other player. Cooperative choice benefits the other player, whereas the defective choice hurts the other player. Given the particular payoffs in Prisoner's Dilemma games, it is *always* rational to defect on the other player as long as the game is one-shot and not repeated infinitely. Regardless of what the other player does, you get a higher payoff by defecting than cooperating.

In one-shot games, there is nothing the other player can do to punish you for defecting. When the game is repeated infinitely, however, it becomes rational for you to cooperate in a Prisoner's Dilemma game because then the other player can punish you on future rounds for defecting. But there are no such concerns for possible future retaliation in a one-shot game. In typical experimental settings for Prisoner's Dilemma games, the two players interact via computers and never face each other in person. In addition, the researchers make sure that the two players will never run into each other before, during and after the experiment, so the two players will remain completely anonymous to each other. Under these experimental conditions, it is always rational to defect on the other player and receive higher payoffs. There are no negative consequences for defection.

Yet experiment after experiment conducted in the last half century show that roughly half of the players of one-shot Prisoner's Dilemma games make the theoretically irrational decision to cooperate.[18] This has been one of the longstanding unsolved mysteries in game theory for more than 50 years. There are some ideas, but no one yet knows for sure exactly why half the people make the irrational decision to cooperate in one-shot Prisoner's Dilemma games in clear contradiction to the prediction of their elegant mathematical models. Microeconomics assumes that all human actors are rational, yet the evidence from these experiments seems to suggest that half of them are not. And microeconomics cannot explain why.

From the perspective of the Savanna Principle, it may be because the two conditions that are theoretically necessary for the prediction of universal defection in one-shot Prisoner's Dilemma games—complete anonymity and noniteration—did not exist in the ancestral environment. There was no such thing as anonymous exchange in the ancestral environment because there were no computer-mediated interactions that would make it possible; all exchanges and interactions in the ancestral environment were face-to-face. And very few, if any, social exchanges were one-shot. Our ancestors lived in a small band of about 150 related individuals all their lives. Everyone in their band was a relative, friend, or ally for life.[19]

So the Savanna Principle suggests that the human brain may have difficulty truly comprehending one-shot games and completely anonymous exchanges, because there were no such things in the ancestral environment. As a result, some individuals may act as if the anonymous one-shot games are face-to-face repeated games, the only kind that existed in the ancestral environment, and decide to cooperate, because it *is* rational to cooperate in nonanonymous repeated games.

This may be why as many as half the people in one-shot Prisoner's Dilemma games make the irrational choice to cooperate. Further, the Savanna-IQ Interaction Hypothesis, an idea that I introduce in Chapter 4, can potentially explain *which 50%* of the people are likely to cooperate in one-shot Prisoner's Dilemma games.

When Inclusion Costs and Ostracism Pays, Ostracism Still Hurts

An incredibly ingenious experiment recently conducted by a couple of social psychologists provides yet another illustration of the Savanna Principle in operation.[20]

Humans are a highly social species, and they rely and depend on each other for survival. For this reason, humans have always lived in social groups. Because humans are highly dependent on others in their groups, ostracism—being excluded from their social groups and the benefits they provide—has always been costly throughout human evolutionary history, and their very survival has often depended on being included in their groups.

It is therefore not surprising at all that humans have evolved psychological mechanisms that incline them to seek group affiliation and avoid ostracism. Studies examining the human brain using the fMRI (functional magnetic resonance imaging) technology have revealed that being ostracized activates the same region of the brain that lights up when individuals experience physical pain.[21] In other words, humans are designed to *feel physical pain* when they are ostracized. Given how dangerous being excluded from the group is for human survival and how very costly ostracism is, especially in the ancestral environment of the African savanna, this makes perfect evolutionary sense. Our ancestors who did not mind being ostracized and didn't feel any pain about it probably didn't live long enough to produce many children.

But what if ostracism was not costly at all? What if, instead, being included is costly and being excluded is beneficial? Would people then come to enjoy being excluded and fear being included? This is the question that motivated Ilja van Beest and Kipling D. Williams to conduct their ingenious experiment in their 2006 article "When Inclusion Costs and Ostracism Pays, Ostracism Still Hurts" published in the premier journal in social psychology, *Journal of Personality and Social Psychology.*

In their experiment, van Beest and Williams use a variant of a multi-player computer game called Cyberball. An individual plays Cyberball with two other players. Each player can see the other two players on the computer screen, but they are in other locations; each player is alone in the room. The game is very simple. The three players toss a ball on the computer screen back and forth with each other. If you get the ball, you can toss it to either of the other two players, and whoever receives the ball tosses it to one of the other players. Each player has a choice of two players to toss the ball to.

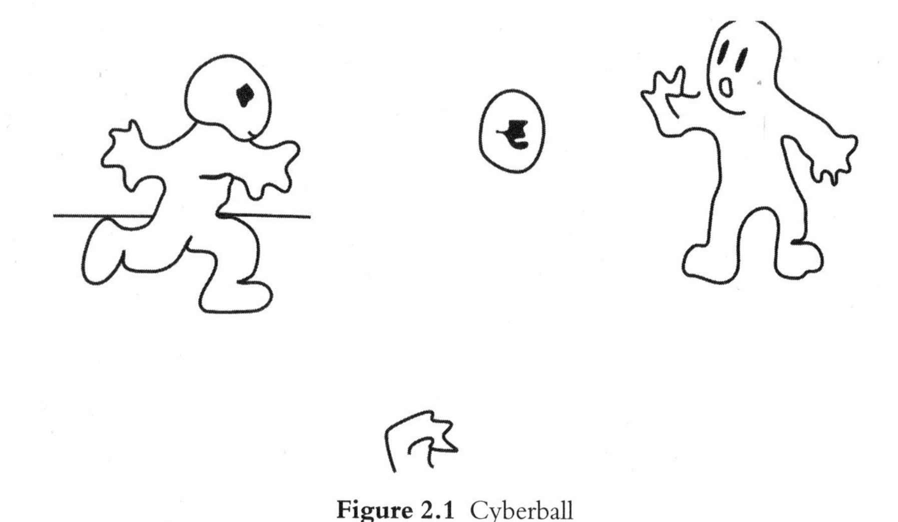

Figure 2.1 Cyberball

Figure 2.1 is a screenshot from Cyberball (courtesy of Kipling D. Williams). If you are a player in this game, the left hand that you see at the bottom of the screen is yours, and you see the other two players in the game in front of you. In the screen shot, you are observing one of the other players (on the left) throwing the ball to the other player (on the right). You are therefore *not* involved in this particular ball toss. The player on the left has chosen the other player, not you, to whom to toss the ball.

Unbeknownst to the human player, however, the other two players on the screen are simulated actors programmed by the researchers to behave in certain ways. The experiment has a 2 (inclusion vs. exclusion) x 2 (gain vs. loss) design. In some games, the human player is included in a fair share of the ball tosses. This is the "inclusion" condition. In other games, after a couple of tosses at the beginning, the human player is completely excluded from the ball toss, and watches the other two players toss the ball back and forth with each other, completely ignoring and excluding the human player. This is the "exclusion" condition.

In some games, in both "inclusion" and "exclusion" conditions, the human players earns 50 cents every time they touch the ball (when they are included in the ball toss). This is the "gain" or "inclusion pays" condition. In other games, in both the "inclusion" and "exclusion" conditions, the human players *lose* 50 cents every time they touch the ball; in other words, in this condition, people are financially better off if they are excluded from the ball tosses. This is the "loss" or "exclusion pays" condition.

Van Beest and Williams's experimental design makes these two factors completely independent of each other. Some subjects gain money while being included, some subjects gain money while being excluded. Other subjects lose money while being included, still others lose money while being excluded. Then, after the Cyberball game is over, the researchers measure the subjects' satisfaction and mood.

It makes perfect sense that human players who were excluded from the ball toss in the "inclusion pays" condition were hurt by being excluded. They would have earned more money if they were included in the game, but they were not, so they felt hurt. No surprises here. What *is* surprising is that people in the "exclusion pays" condition were also hurt when they were excluded. These people made more money by being excluded from the game, yet they were equally hurt by not being included in the ball toss by the other two players. How could this be? How could people feel hurt when they were doing better?

The Savanna Principle can offer one potential answer. Throughout the course of human evolution, exclusion was *always* costly and inclusion was *always* beneficial. These two things always went together, because there were no evil experimental psychologists in the ancestral environment to manipulate these variables independently. There were no such things as "beneficial exclusion" and "costly inclusion." Our ancestors were never in the "exclusion pays" condition. The human brain therefore cannot comprehend such a thing. The human brain implicitly and unconsciously assumes that *all* ostracism is costly, just as it assumes that all realistic images of people whom they see on a regular basis (and who don't try to kill or harm them in any way) are their friends, even when these people are on TV.

Microeconomic theory, or any other theory of human behavior which assumes that human behavior is rational and based on carefully calculated cost-benefit analysis, cannot explain van Beest and Williams's remarkable findings that humans are happy to lose money and sad to make money. Without the Savanna Principle, it would be difficult to explain why ostracism makes people sad when it pays. This is one of the many reasons why evolutionary psychology is superior to microeconomics (or any other theory) as an explanation for human behavior, even when we are not talking about sex differences.[22]

The fundamental insight of evolutionary psychology, expressed in the Savanna Principle, is that the human brain responds to the environment as if it were still the African savanna in the ancestral environment (for the most part, during the Pleistocene Epoch, 1.6 million to 10,000 years ago). *You* the person may consciously know that this is the 21st century, and you are a stockbroker in New York, an artist in Seattle, a housewife in San Francisco, or a student in Kansas City, *but your brain doesn't know that.* Your brain, unconsciously and implicitly, still thinks that you are a hunter-gatherer living on the African savanna more than 10,000 years ago, where there was no TV or psychology experiments or virtually anything else you see around you today. As you can imagine, implications of this fundamental observation of evolutionary psychology for our modern life are significant and widespread.

Chapter 3

What Is Intelligence?

Intelligence (or, more precisely, *general intelligence*) refers to the ability to reason deductively or inductively, think abstractly, use analogies, synthesize information, and apply it to new domains.[1] Perhaps no other concept in science suffers from greater misunderstanding and is plagued with more misconceptions than the concept of intelligence. Many of these misconceptions are politically motivated by the equation of intelligence with human worth that I mention in the Introduction. Before I discuss how general intelligence evolved and what it is good for, I want to attempt to dispel these common misconceptions about intelligence.[2]

Common Misconceptions about Intelligence

Misconception 1: IQ Tests Are Culturally Biased

Probably the most pervasive misconception about intelligence is that IQ tests, which measure intelligence, are culturally biased against certain racial and ethnic groups or social classes. This misconception stems from the well-established and replicated fact that different racial and ethnic groups on average score differently on standardized IQ tests. As I mention in the Introduction, social scientists and lay public alike *assume* (without any logical or empirical support) that everyone (and all racial and ethnic groups) is equally intelligent because they are all equally worthy human beings. If everybody is equally intelligent, yet some groups consistently score lower than others, then, the argument goes, it must by definition mean that the IQ tests are culturally biased against the groups who score lower.

But the claim of cultural bias rests *entirely* on the *conviction* and *unquestioned assumption* that everybody and all groups are equally intelligent, which in turn rests *entirely* on the *conviction* and *unquestioned assumption* that intelligence is the ultimate measure of human worth. The claim is untenable once we dismiss these untested and hence *religiously held* convictions and assumptions.

Think about the sphygmomanometer, for a moment. It is the device that doctors and nurses commonly use to measure blood pressure, with the inflatable cuff with Velcro and a mercury manometer to measure the pressure of blood flow. It is an unbiased and accurate (albeit imperfect) device to measure blood pressure. (It is *very* imperfect, as I discuss later in the chapter.) Nobody would argue that it is culturally biased against any racial or ethnic group. Yet there are well-established race differences in blood pressure; blacks on average have higher blood pressure than whites.[3] Does that mean that the sphygmomanometer is culturally biased against (or for!) blacks? Is blood pressure a racist concept?

Of course not. It simply means that blacks on average have higher blood pressure than whites. Nothing more, nothing less.

Or think of the bathroom scale. Once again, the bathroom scale is an unbiased and accurate (albeit imperfect) device to measure someone's weight. Nobody would argue that it is culturally biased against certain groups. Yet on any bathroom scale, women on average "score" lower than men, and Asians on average "score" lower than Caucasians. Does that mean the bathroom scale is culturally biased against women or Asians? Is weight a sexist or racist concept? Of course not. It simply means that women on average weigh less than men, and Asians on average weigh less than Caucasians. Nothing more, nothing less.

Nobody argues that blood pressure is a racist concept, or that the sphygmomanometer is culturally biased, because nobody equates (low) blood pressure with human worth. As a result, nobody gets upset about observed race differences in blood pressure. Nobody argues that weight is a racist or sexist concept, or that the bathroom scale is culturally biased, because nobody equates weight with human worth. As a result, nobody gets upset about observed sex or race differences in weight. Why are race and sex differences bona fide evidence of bias *only* with IQ tests?

The single most accurate IQ test currently available is called the Raven's Progressive Matrices. Intelligence researchers universally consider it to be the single best test to measure general intelligence because scores on this test are more strongly correlated with the underlying dimension of general intelligence than any other single intelligence test. (In technical language, Raven's Progressive Matrices are more highly *g*-loaded than any other cognitive test.) The test comes in three different versions: the standard version (Raven's Standard Progressive Matrices); the advanced version for college students and other more intelligent people (Raven's Advanced Progressive Matrices), designed to discriminate the higher end of the IQ distribution more precisely; and the multi-color version for children (Raven's Colored Progressive Matrices).

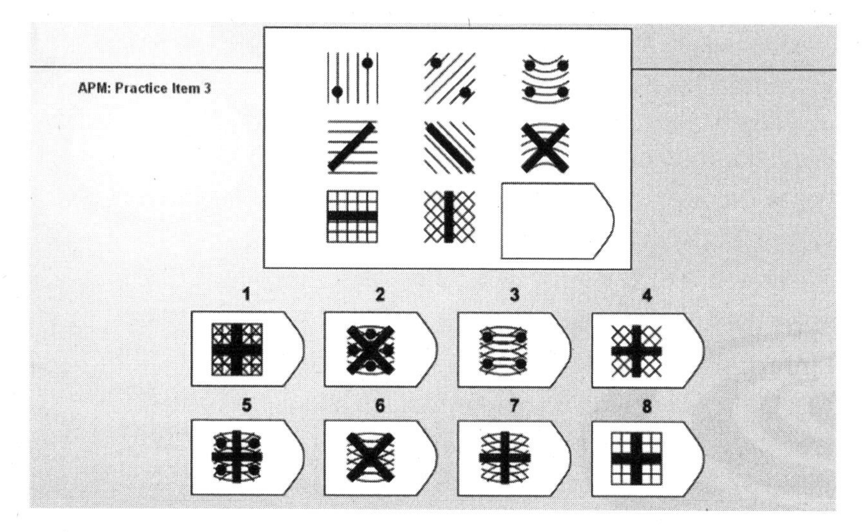

Figure 3.1 A question from the Raven's Advanced Progressive Matrices

Here is an example of a question item from Raven's Advanced Progressive Matrices. The test comes with only one instruction: Choose the figure that fits the next in the progression of matrices. Which one of the eight alternatives comes next?

All question items in all versions of Raven's Progressive Matrices are very similar to this one. Can anyone tell me exactly how this question, and all the other similar questions that comprise the Raven's Progressive Matrices, can possibly be culturally biased against any group? The question is a pure measure of reasoning ability. The only thing it's biased against is the inability to think logically.

By the way, if you are wondering, the correct answer to the above question is 7.

Misconception 2: Nobody Knows What Intelligence Is, because Intelligence and IQ Are Not the Same Thing

A related misconception that people have is the claim that IQ is not a measure of general intelligence. Some people believe in the concept of intelligence; they know that some people are more intelligent than others. But they do not believe that IQ tests accurately measure individuals' intelligence, once again, because IQ test scores typically show average differences between different groups and they believe that individuals from different groups on average *must be* equally intelligent.

Contrary to this view, intelligence researchers unanimously agree that intelligence is *exactly* what IQ tests measure,[4] in the same way that your weight is exactly what your bathroom scale measures. To maintain that intelligence is real and some people are more intelligent than others, yet IQ tests do not accurately measure intelligence is akin to claiming that weight is real and some people are heavier than others, but the bathroom scale does not accurately measure weight. It simply doesn't make any sense.

I have just said that Raven's Progressive Matrices is the *single* best IQ test currently available, and that is true. But there is actually a better way to measure someone's general intelligence than Raven's, and that is to administer a *series* of different cognitive tests. The best way to assess someone's level of general intelligence is to administer a large number of cognitive tests like vocabulary, verbal comprehension, arithmetic, digit span (to measure the ability to repeat a sequence of digits after it is given, sometimes exactly as it is given, sometimes backwards), spatiovisual rotation (to measure the ability to imagine what a three-dimensional object would look like if it is rotated in space), etc. You will recall from the Introduction that this is precisely how NCDS measures intelligence, which is why NCDS has one of the best measures of general intelligence of all large-scale national surveys.

Across individuals, performances on all these cognitive tests are highly positively correlated. In other words, people who do well in verbal comprehension tests tend also to do well on arithmetic tests, and they have better ability to visualize a three-dimensional object from a different angle or to repeat a sequence of digits that is given to them backwards. Contrary to popular belief, people who are good with concrete tasks are also good with abstract tasks; people who are good with numbers are also good with words.

For example, in a classic paper published in 1904, Charles Spearman shows that students' relative school performance in mathematics is highly correlated with their performance in classics ($r = .87$), French ($r = .83$), English ($r = .78$), pitch discrimination ($r = .66$) and music ($r = .63$). (The "r" is a measure of association between two variables, known in statistics as the *correlation coefficient*. It varies from -1, when the two variables are perfectly negatively correlated, through 0 when they are completely unrelated to each other, to +1 when they are perfectly positively correlated. As you can see, all of the correlations reported by Spearman are very highly positive.) In fact, the students' relative performance in music is more highly correlated with their mathematical ability than with their pitch discrimination ($r = .40$)!

In the NCDS data, at age 16, the correlation between verbal comprehension and mathematical comprehension is .654, which once again is very high. As I note below in the next section, the correlation between true blood pressure and blood pressure measured by the sphygmomanometer is about .50. It means that using a verbal comprehension test to measure one's mathematical ability, or using one's relative performance in mathematics to measure one's relative performance in musical ability, is more accurate than using the sphygmomanometer to measure blood pressure. That is how high all measures of cognitive abilities are intercorrelated.

At the same time, it is also important to remember that as highly as verbal comprehension and mathematical comprehension are correlated in NCDS ($r = .654$), one can explain less than half of the variance in the other. ("Explained variance" in one variable by another is computed by squaring the correlation coefficient between them. So it means that one's score in the verbal comprehension test explains 43% ($(.654)^2 = .428$) of the variance in mathematical comprehension and vice versa.) It means more than half of the variance in mathematical comprehension test scores across individuals cannot be explained by their scores in the verbal comprehension test scores.

What psychometricians (whose job it is to measure intelligence accurately and devise tests to do so) do then is to subject individual scores on all these cognitive tests to a statistical technique called *factor analysis.* What factor analysis does is to analyze the correlations between all pairs of cognitive tests and then measure an individual's *latent* cognitive ability that underlies their performance on all of the cognitive tests. This latent cognitive ability is general intelligence. Factor analysis also eliminates all random measurement errors that are inevitably associated with any individual cognitive test as a measure of intelligence. So it can measure general intelligence purely, without any random measurement errors.

The IQ score thus obtained is a pure measure of intelligence.[5] It measures someone's ability to think and reason in various contexts and situations, such as numerical manipulations like arithmetic, verbal comprehension like reading, and mental visualization like spatiovisual rotation. Believe it or not, all these cognitive abilities have something in common, and that something is general intelligence. So intelligence is *precisely* and *exactly* what IQ tests measure. Intelligence is what allows us to perform on all kinds of cognitive tests.

Misconception 3: IQ Tests Are Unreliable

Unlike other misconceptions about intelligence, there is some truth to this one, in the sense that IQ tests are not *perfectly* reliable. IQ tests have some measurement errors, which is why psychometricians perform factor analysis to eliminate such random errors in measurement. So it is true that IQ tests are not *perfectly* reliable, but then no scientific measurements ever are.

If the same individuals take different IQ tests on different days, or even on the same day, their scores will be slightly different from test to test (but only slightly). So IQ tests do not give the perfect measurement of someone's intelligence. But then, if you step on the bathroom scale, get the reading, step off, and step on it again, it will give you slightly different readings as well. The same is true if you measure your height, your shoe size, and your vision. No measurements of any human quality are perfectly reliable.

So the measurement of intelligence is no different from the measurement of any other human trait. But nobody ever claims that, because the measurement of weight is never perfectly reliable, there is no such thing as weight and weight is a culturally constructed concept. But that's exactly what people who are unfamiliar with the latest psychometric research think about intelligence. Intelligence is no less real than height or weight, and its measurement is just as reliable (or unreliable).

In fact, Arthur R. Jensen, probably the greatest living intelligence researcher, claims that IQ tests have higher reliability than the measurement of height and weight in a doctor's office.[6] He says that the reliability of IQ tests is between .90 and .99 (meaning that random measurement error is between 1% and 10%), whereas the measurement of blood pressure, blood cholesterol, and diagnosis based on chest X-rays typically has a reliability of around .50.

Reliability is the correlation coefficient between repeated measurements. If the measurement instrument is unbiased (as IQ tests are as a measure of general intelligence and the sphygmomanometer is as a measure of blood pressure), then the reliability translates into the correlation coefficient between the true values and the measured values. The reliability of .50, for example, like the reliability of the sphygmomanometer as a measure of blood pressure, means that the correlation between individuals' true blood pressure and the readings on the sphygmomanometer is only .50. In contrast, the reliability of .90-.99, for example, the reliability of IQ tests as a measure of general intelligence, means that the correlation between individuals' general intelligence and their IQ test scores is .90–99. So the measurement of intelligence is nearly twice as accurate as the measurement of blood pressure, yet nobody ever claims that blood pressure is not real or that it is a culturally constructed concept.

Misconception 4: Genes Don't Determine Intelligence, Only the Environment (Education and Socialization) Does

This is another widely held misconception about intelligence. It is true that genes don't determine intelligence *completely*; they only do so substantially and profoundly.

Heritability is the measure of the influence of genes on any trait.[7] Heritability of 1.0 means that genes determine the traits completely and the environment has absolutely no effect. As I mention in Chapter 1, some genetic diseases like Huntington's disease have a heritability of 1.0; genes entirely determine whether or not you will get Huntington's disease. If you have the affected genes for the disease, it does not matter at all how you live your life or what your environment is; you will develop the disease. One's natural eye color or natural hair color also has a heritability of 1.0. So does one's blood type. Very few other human traits have a heritability of 1.0.

On the other hand, a heritability of 0 means that genes have absolutely no influence on the given trait, and the environment completely determines whether or not someone has the trait. *No human traits have a heritability of 0; genes partially influence all human traits to some degree.* (This is known as Turkheimer's first law of behavior genetics.[8])

Most personality traits and other characteristics—like whether you are politically liberal or conservative or how likely you are to get a divorce—have a heritability of .50; they are about 50% determined by genes.[9] In fact, most personality traits and social attitudes follow what I call the 50–0–50 rule:[10] roughly 50% heritable (the influence of genes), roughly 0% what behavior geneticists call "shared environment" (parenting and everything else that happens within the family to make siblings similar to each other), and roughly 50% "nonshared environment" (everything that happens outside of the family to make siblings different from each other). It turns out that parenting has very little influence on how children turn out.[11]

Of course, this emphatically does *not* mean that parents are not important for how children turn out; they are *massively* and *supremely* important because children get their genes from their genetic parents. It simply means that *parenting*—how parents raise their children—is unimportant. This is why adopted children usually grow up to be nothing like their adoptive parents who raised them and a lot like their biological parents (or their twin reared apart) whom they have never even met.[12]

One of the very few exceptions to the 50–0–50 rule is intelligence, for which heritability is *larger.* Heritability of general intelligence increases from about .40 in childhood to about .80 in adulthood. Among adults, intelligence is about 80% determined by genes.

Yes, heritability of intelligence *increases* over the life course, and genes become *more* important as one gets older. This may at first sight seem counterintuitive, but it really isn't. This is because for adults the environment is part of their genetic makeup whereas for children it isn't. Children must live in the environment created by their parents, older siblings, teachers, neighbors, clergy, and other adults. In contrast, adults determine their own environment to a much greater extent than children do. So for adults, genes and the environment become more or less the same thing, whereas for children they are not. For adults, when the environment influences their intelligence, it shows up as the influence of their genes, which largely determine their environment, whereas for children it does not. This is why the influence of genes increases dramatically throughout life.

Sorry, Education Does *Not* Increase Your Intelligence; It's the Other Way Around

A subcategory of this common misconception is that you can *become* more intelligent, by reading more books, attending better schools, or receiving more education. It is true that there are strong associations among these traits. People who read more books *are* more intelligent; people who attend better schools *are* more intelligent; and people who attain more education *are* more intelligent. But the causal order is the opposite of what many people assume. There are associations among these traits, because more intelligent people read more books, attend better schools (partly because their parents are more intelligent and therefore make more money), and receive more education.

Early childhood experiences do affect adult intelligence, but they mostly function to *decrease* adult intelligence, not to increase it. Childhood illnesses, injuries, malnutrition, and other adverse conditions influence adult intelligence negatively, and these individuals often fail to fulfill their genetic potential. But there are very few childhood experiences that will *increase* adult intelligence much more than their genes would have inclined them to have.

Somewhat paradoxically, the wealthier, the safer, and the more egalitarian the nations become, the *more* (not less) important the genes become in determining adult intelligence. In poor nations, there are many children who grow up ill, injured or malnourished, and these children will decrease the correlation between genes and adult intelligence. In wealthy societies like the United States, where very few children now grow up ill and malnourished, the environment is more or less equalized. When the environment becomes equal for all individuals, it has the same effect for everyone and it can no longer explain any variance in the individual outcome. (Statistically, a factor that does not vary between individuals cannot be correlated with individual differences in an outcome. And no correlation means no explained variance, as zero squared equals zero.) So the more equal the environment between individuals, the more important the influence of genes becomes. A longitudinal study of Scottish people born in 1921 and 1936 shows that their intelligence does not change much after the age of 11.[13] Their intelligence at age 11 is very strongly correlated with their intelligence at age 80.

So contrary to the popular misconception, genes largely (though, even for adults, never completely) determine intelligence. In fact, intelligence is one of the most heritable of all human traits and characteristics. For example, intelligence is just as heritable as height.[14] Everybody knows that tall parents beget tall children, and nobody ever questions the strong influence of genes on height, yet they vehemently deny *any* influence of genes on intelligence.

There is something curious about heritability. A trait's heritability and its adaptiveness (how important it is for survival and reproductive success) are generally inversely related: The more adaptive the trait is (the more important it is for the organism's survival and reproductive success), the *less* heritable it is.[15] This is because, when a trait is crucial for survival and reproductive success, every individual must have it at the optimal and most efficient level. Evolution cannot "allow" it to vary across individuals. It is only when a trait is less important for survival and reproductive success that evolution can "allow" it to vary across individuals. Thus, according to the basic principles of quantitative genetics, the fact that general intelligence is highly heritable suggests that it is not very important for our survival and reproductive success, as I argue throughout the book.

How Did General Intelligence Evolve?

As I discuss in Chapter 1, evolutionary psychology contends that the human mind consists of evolved psychological mechanisms. Evolved psychological mechanisms are *domain-specific.* It means that these evolved and innate solutions to adaptive problems each operate only in their own specific narrow domains of life.

For example, the cheater detection mechanism, which was among the very first evolved psychological mechanisms to be discovered,[16] operates only in the domain of social exchange; it helps us detect potential cheaters when they try to cheat us out of a fair exchange. But the cheater detection mechanism does not help us, nor is it operative, in any other domains of life. It does not help us, for example, learn our native language. It does not help us decide how to allocate limited parental resources among our children (in other words, which of our children to favor unconsciously—yes, parents do have favorites among their children[17]). And it does not help us recognize familiar faces of the people we know.

By the same token, the language acquisition device, which helps us learn our native language from our mothers when we are small children,[18] does not help us detect cheaters in social exchange, decide how to allocate parental resources, or recognize familiar faces. In fact, it does not help us do *anything* except for the one task of learning our native language. It does not even help us learn second and third languages as adults, which is why learning a foreign language is so much more difficult than learning to speak our native language as a child, which comes very naturally and easily to us (because we have the innate ability to do so).

Evolved psychological mechanisms are domain-specific because adaptive problems, which they are designed to solve, are domain-specific. All problems of survival and reproduction happen in specific domains; there were no general problems that did not happen in a specific context for our ancestors to solve.

Evolution did not give us a domain-general solution like a computer because there were no domain-general problems like an IQ test in the ancestral environment.

However, if the contents of the human brain are domain-specific, how can evolutionary psychology explain general intelligence, which is seemingly domain-general, not domain-specific? Isn't general intelligence a domain-general solution?

General intelligence thus posed a significant theoretical problem for evolutionary psychology for a long time. How can evolutionary psychology explain the evolution of general intelligence? I believe that what is now known as general intelligence may have originally evolved as a domain-specific adaptation to deal with evolutionarily novel, nonrecurrent problems.[19] Now what does that mean?

The Pleistocene Epoch (between 1.6 million and 10,000 years ago), during which humans evolved, was a period of extraordinary constancy and continuity. Nothing much happened for more than a million years. Our ancestors were hunter-gatherers on the African savanna all of their lives. Their grandparents were hunter-gatherers on the African savanna all their lives. Their parents were hunter-gatherers on the African savanna all their lives. Their children were hunter-gatherers on the African savanna all their lives. Their grandchildren were hunter-gatherers on the African savanna all their lives.

This kind of constancy is difficult for us to fathom. In our grandparents' generation, most people were farmers; now very few people are. In our parents' generation, most people were factory workers; now very few people are. Now most of us, regardless of our specific occupation, conduct our business and trade at least partly on our computer, a device that did not exist in our grandparents' or even our parents' generation.

It is against the backdrop of the extreme stability of our ancestors' environment during the Pleistocene Epoch that all of our psychological adaptations evolved. For instance, those who had a taste for sweet and fatty food during the Pleistocene Epoch lived longer and reproduced more successfully, by acquiring more calories, than those who did not have such a taste for sweet or fatty food.[20] Or those who preferred a certain landscape for their habitat lived longer and reproduced more successfully, by avoiding potential predators in hiding, than those who did not have such a preference.[21] The evolution of psychological mechanisms—or any adaptation, physical or psychological—assumes a stable environment. Because evolution usually takes place very, very slowly, over tens and hundreds of thousands of years, solutions cannot evolve in the form of psychological mechanisms if the problems keep changing.[22] The fact that we have so many evolved psychological mechanisms today is testimony to the extraordinary stability and constancy of the ancestral environment.

Technically, the speed of evolution depends on the strength of *selection pressure*—how crucial it is for survival and reproduction to solve a given adaptive problem. The rate of evolution of a trait is proportional to the adaptiveness of the trait—the correlation between possessing the trait and being able to reproduce. For example, if cosmic rays from an explosion of a nearby supernova render all but redheads sterile, then in just one generation everyone on earth will be a redhead because no one else will be able to reproduce. But selection pressures are usually much weaker (for example, the cosmic rays will allow redheads to reproduce at a 5% greater rate than everyone else), so evolution of most traits take many, many generations.

Because adaptive problems in the ancestral environment remained more or less the same generation after generation, our evolved psychological mechanisms were sufficient for our ancestors to solve them. In this sense, our ancestors did not really have to *think* in order to solve their adaptive problems. They didn't have to think, for instance, what was good to eat. All they had to do was to eat and keep eating what tasted good to them (sweet and fatty foods that contained high calories), and they lived long and remained healthy. People who preferred the wrong kind of food—like brightly colored mushrooms or an entirely vegetarian diet—died off before leaving too many offspring, and we did not inherit our psychological mechanisms from them. All the thinking had already been done by evolution, so to speak, which then equipped our ancestors with the correct solutions in the form of innate domain-specific psychological mechanisms. For the most part, our ancestors never had to figure out problems on their own.

Even in the extreme continuity and constancy of the ancestral environment, however, there were likely occasional problems that were evolutionarily novel and nonrecurrent, which required our ancestors to think and reason in order to solve. Such problems may have included, for example:

1. Lightning has struck a tree near the camp and set it on fire. The fire is now spreading to the dry underbrush. What should I do? How could I stop the spread of the fire? How could I and my family escape it? (Since, as they say, lightning never strikes the same place twice, this is guaranteed to be a nonrecurrent problem!)
2. We are in the middle of the severest drought in as long as anyone can remember. Nuts and berries at our normal places of gathering, which are usually plentiful, are not growing at all, and animals are scarce as well. We are running out of food because none of our normal sources of food are working. What else can we eat? What else is safe to eat? How else can we procure food?
3. A flash flood has caused the river to swell to several times its normal width, and I am trapped on one side of it while my entire band is on the other side. It is imperative that I rejoin them soon. How could I cross the rapid river? Should I walk across it? Or should I construct some sort of buoyant vehicle to use to get across it? If so, what kind of material should I use? Wood? Stones?

To the extent that these evolutionarily novel, nonrecurrent problems happened frequently enough in the ancestral environment (a different problem each time[23]) and had serious enough consequences for survival and reproduction, then any genetic mutation that allowed its carriers to think and reason would have been selected for, and what we now call "general intelligence" could have evolved as a domain-specific psychological mechanism for the domain of evolutionarily novel, nonrecurrent problems, which did not exist in the ancestral environment and which there are no dedicated psychological modules to solve.

From this perspective, general intelligence may have become universally important in modern life[24] only because our current environment is almost entirely evolutionarily novel. In the ancestral environment, general intelligence might have been no more important than any other domain-specific evolved psychological mechanism, like the cheater detection mechanism or the language acquisition device. General intelligence helped our ancestors only in the very narrow domain of evolutionary novelty—evolutionarily novel problems were by definition few and far between in the ancestral environment—just as the cheater detection mechanism helped them only in the very narrow domain of social exchange and the language acquisition device helped them only in the very narrow domain of the native language acquisition. General intelligence became more important than other evolved psychological mechanisms only because our environment has changed so radically in the last 10,000 years and most of the problems we face today are evolutionarily novel. The importance of general intelligence may itself be evolutionarily novel.

This theory suggests that more intelligent individuals are better than less intelligent individuals at solving problems *only if* they are evolutionarily novel. More intelligent individuals are *not better* than less intelligent individuals at solving evolutionarily familiar problems that our ancestors routinely had to solve. The theory suggests that the performance of general intelligence, as but one domain-specific psychological mechanism, is independent of the performances of all the other domain-specific psychological mechanisms. I review some of the evidence for this theory of the evolution of general intelligence in the next chapter.

Cognitive Classes

Before we leave this chapter on intelligence, I'd like to introduce the concept of "cognitive classes," which I use frequently throughout the rest of the book. It is a way of grouping individuals into five ordinal categories by their intelligence, from highest to lowest. This is a classification system that other scholars have invented[25] and I have used before in my own work.[26] The labels for the cognitive classes are used merely as a convenient shorthand, without any connotations.

Very bright (IQ > 125: roughly 5% of the US population)
Bright (110 < IQ < 125: roughly 20% of the US population)
Normal (90 < IQ 110: roughly 50% of the US population)
Dull (75 < IQ < 90: roughly 20% of the US population)
Very dull (75 < IQ: roughly 5% of the US population)

Here is a way to give you a quick flavor of what these cognitive classes mean. Among white Americans, 75% of those who earn a bachelor's degree are "very bright"; none are "very dull." In contrast, 64% of those who drop out of high school are "very dull"; none are "very bright."[27]

Chapter 4

When Intelligence Matters (and When It Doesn't)

The Savanna-IQ Interaction Hypothesis

The logical intersection of the Savanna Principle—discussed in Chapter 2—and the theory of the evolution of general intelligence—discussed in Chapter 3—suggests a qualification of the Savanna Principle. If general intelligence evolved to deal with evolutionarily novel problems, then the human brain's difficulty in comprehending and dealing with entities and situations that did not exist in the ancestral environment (proposed in the Savanna Principle) should interact with general intelligence. In other words, the Savanna Principle should hold stronger among less intelligent individuals than among more intelligent individuals. More intelligent individuals should be better able to comprehend and deal with evolutionarily novel (but *not* evolutionarily familiar) entities and situations than less intelligent individuals.

So I now propose the Savanna-IQ Interaction Hypothesis[1] (or "the Hypothesis"). The Hypothesis qualifies and elaborates on the Savanna Principle by introducing intelligence and how it modifies the operation of the Savanna Principle.

> *The Savanna-IQ Interaction Hypothesis: Less intelligent individuals have greater difficulty comprehending and dealing with evolutionarily novel entities and situations that did not exist in the ancestral environment than more intelligent individuals. In contrast, general intelligence does not affect individuals' ability to comprehend and deal with evolutionarily familiar entities and situations that existed in the ancestral environment.*

Recall the definition of "comprehension" from Chapter 2 as the true, logical, and scientifically and empirically accurate understanding of how something works. Now I am going to review the empirical evidence for the Savanna-IQ Interaction Hypothesis in many different domains of life.

Back to TV Friends

In Chapter 2, I discuss my 2002 study,[2] which suggests that people may have some implicit difficulty distinguishing their "TV friends"—characters that they repeatedly see on TV—from their real friends. The more they watch certain types of TV shows, the more satisfied they become with their friendships, just as they do if they have more friends or socialize with them more frequently. This makes perfect sense from the perspective of the Savanna Principle. Because there were no realistic electronic (or photographic) images of other humans in the ancestral environment, the human brain has difficulty with such images. Since all realistic images of other humans in the ancestral environment *were* other humans, the human brain implicitly assumes that any such images of other humans, who don't attempt to kill or maim them (which very few TV characters do), are their friends.

In 2006, after I formulated the initial ideas behind the Hypothesis, I reanalyzed the same data from the General Social Surveys to see if individuals' IQ was related to their implicit tendency to confuse "TV friends" and real friends.[3] And it was. This tendency, which I initially thought (according to the Savanna Principle) was a universal human trait in 2002, appears to be limited to men and women below median intelligence (consistent with the Savanna-IQ Interaction Hypothesis). Those who are above median in intelligence do not report greater satisfaction with friendships as a function of watching more TV; only those below median intelligence do.

This seems to suggest that the evolutionary constraints on the brain predicted by the Savanna Principle, whereby individuals have implicit difficulty with recognizing realistic electronic images on TV for what they are, appear to be weaker or altogether absent among more intelligent individuals. Since truly enjoying the experience of watching TV requires suspension of disbelief and *not* really understanding that characters repeatedly seen on the screen are highly paid actors hired to play scripted roles, this new finding can potentially explain why less intelligent individuals tend to enjoy the experience of watching TV more than more intelligent individuals do.

Of course, just like everything else I say in this book, the negative association between general intelligence and the enjoyment of TV is an empirical generalization, for which there are many exceptions. I personally happen to love watching TV myself. However, among my highly intelligent academic colleagues, there are many who do not watch television at all. Some don't even own a television set. I'm often frustrated with them, because I have nothing to talk to them about if they don't watch television at all. I share no common points of reference with such people. Like the comedy writer Tina Fey, I believe that American television is one of the greatest things about the country, and I feel sorry for people who cannot appreciate its many wonderful programs. Nevertheless, the Savanna-IQ Interaction Hypothesis can explain why many intelligent people do not watch television at all.

We all know people who have a tendency to talk back and speak to characters they see while they are watching TV or movies. I believe this habit also stems from their implicit confusion of "TV friends" and real friends, the confusion of realistic electronic images of other human beings on the screen and real human beings. I would therefore predict that less intelligent individuals are more likely to talk back and speak to TV and movie screens, and further that these are precisely the people who enjoy watching TV and movies more than others do. Less intelligent individuals are less likely truly to comprehend electronic images on the screen and more likely to forget unconsciously that they are not real live human beings who can hear us when we talk back to them.

Back to Pornography

In Chapter 2, I explain that men's and women's brains cannot truly comprehend pornography—realistic photographic and videographic images of sexually aroused men and women—because no such thing existed in the ancestral environment. The Savanna Principle can explain why men and women confuse porn stars with real potential sex partners, just as they confuse "TV friends" with real friends. Now the Hypothesis would predict that such confusion of porn stars with real potential sex partners might interact with general intelligence, just as does the confusion of "TV friends" with real friends.

In the first independent empirical confirmation of the Hypothesis, conducted by researchers other than myself, Gorge A. Romero and Aaron T. Goetz, two young evolutionary psychologists at California State University-Fullerton, demonstrate that this indeed is the case.[4]

In their survey, Romero and Goetz measure three things from their male respondents: (1) their perception of women's sexuality and sexual behavior; (2) their consumption of pornography; and (3) their general intelligence. Their analysis demonstrates that there is a positive association between men's consumption of pornography and how sexually promiscuous they think women are, in other words, how likely they think women in real life behave like porn stars. The more frequently they watch porn, the more they believe that women are sexually more promiscuous (have a large number of sex partners and one-night stands), and enjoy having casual sex, giving oral sex, receiving anal sex, and having threesomes, just as women typically do in porn movies. However, this confusion of real women and porn stars happens *only if* the men are less intelligent (at least one standard deviation below the mean), *not* if they are of average or above-average intelligence (at least one standard deviation above the mean).

This is a brilliant empirical demonstration of the Hypothesis in operation. Men's brains confuse real women and porn stars—thereby believing that real women act like porn stars by being sexually promiscuous and enjoying unusual sexual acts—but only if they are of below-average intelligence. Romero and Goetz further demonstrate that, once they control for general intelligence, the consumption of pornography *alone* does not increase men's tendency toward this confusion. One must simultaneously have below-average intelligence *and* consume a large quantity of pornography in order to confuse real women with porn stars.

The Failures of the Truly Gifted

Perhaps nothing illustrates the operation of the Hypothesis, and the sharp distinction between the evolutionarily novel and evolutionarily familiar domains of life, more clearly than the profile of the truly gifted. The Study of Mathematically Precocious Youth tracks the lives of more than 5,000 individuals who have been identified as truly gifted in the SAT talent search.[5] Most people take the SAT in the last year of high school at age 17. Participants in the talent search take it in the 7th or 8th grade, before the age of 13. If they score within the top .01% (top 1 in 10,000) for their age, by scoring either more than 700 (out of 800) on the SAT mathematical reasoning ability or more then 630 on the SAT verbal reasoning ability, they are included in the study. Their IQ is therefore higher than 155.

The SAT is considered to be a reasonable IQ test, a test of reasoning ability, not of acquired knowledge, while the ACT is more an achievement test of acquired knowledge.[6] According to the ACT's own website, "The ACT is an *achievement* test, measuring what a student has learned in school. The SAT is more of an *aptitude* test, testing reasoning and verbal abilities."[7]

As you might expect, these individuals of extraordinarily high intelligence achieve equally extraordinary success in the evolutionarily novel domains of formal education and paid employment in the capitalist economy. More than half of them (51.7% of men and 54.3% of women) have earned a doctorate (Ph.D., M.D., or J.D.), compared to the population baseline in the US of 1%. An additional 5.3% of them have earned an MBA, all but one of them in the top 10 US programs. Nearly half (45.8%) of them are university professors, engineers, or scientists; an additional 13.6% are in medicine or law. More than a fifth (21.7%) of those in tenure-track positions in the top 50 US universities are already full professors in their early 30s. (It is virtually unheard of for someone to achieve the rank of full professor in their early 30s in *any* university, let alone in a top 50 US university.) More than a third of the men and about a fifth of the women earn more than $100,000 a year in 2003–2004 in their early 30s. (*Yes, in their early 30s!*) Additionally, 17.8% of the men and 4.3% of the women have earned patents, compared to the population baseline in the US of 1%.

No matter how you slice it, there is no question that these individuals in the Study of Mathematically Precocious Youth go on to experience tremendous success in life, measured by their educational achievement, professional careers, income, and creativity. All of these, however, are evolutionarily novel areas of life, which our ancestors did not have more than 10,000 years ago. How do the same individuals, with IQs higher than 155, fare in evolutionarily familiar areas of life?

Mating and parenting are eminently evolutionarily familiar domains of life. Despite the cumbersome interventions of modern inventions (condoms, sperm banks, internet porn), we still mate, pretty much the same way as our ancestors did 10,000 years ago. Sexual courtship today still involves initial visual and chemical attraction, verbal and physical interaction, mutual mate choice based on social status, physical attractiveness, and moral character as clues to good genes and parental abilities, foreplay, copulation, positive or negative reaction to the mate choice by friends and family, etc. And we still have children as our ancestors did then. Children today, as then, are raised by pair-bonded couples, single mothers and their kin, biological mothers and stepfathers, etc. Few other domains of life today are as evolutionarily familiar as marriage and parenting, so the Hypothesis would predict that more intelligent individuals do not fare better than less intelligent individuals in the domains of marriage and parenting.

This indeed appears to be the case. In stark contrast to their stellar successes in education and employment, the participants in the Study of Mathematically Precocious Youth do not do very well in the evolutionarily familiar domains of marriage and parenting. For example, 64.9% of the men and 69.0% of the women remain childless at age 33, compared to the population baseline of 26.4% in the age group 30–34. The majority of parents only have one child. As a result, the mean number of children is .61 for men and .44 for women, compared to the population baseline of 1.59 for women in the age group 30–34. Despite their extraordinarily high general intelligence, these men and women seem to lag behind everyone else in the evolutionarily familiar domains of marriage and parenting.

The fact that truly gifted individuals have no particular advantage (and often disadvantage) in such evolutionarily familiar domains as mating is illustrated by the following exchange between Stephen Hawking and Larry King. Hawking appeared on *Larry King Live Weekend* on Christmas Day 1999, on the eve of the new millennium, when the following exchange took place:

> *Larry King:* What, Professor, puzzles you the most? What do you think about the most?
> *Stephen Hawking:* Women.
> *Larry King:* Welcome aboard.

The Hypothesis would indeed predict that the Lucasian Professor of Mathematics at the University of Cambridge (a post once held by Isaac Newton), who is putatively the most intelligent person in the United Kingdom today and can figure out the origin and the destiny of the universe, has no particular advantage in evolutionarily familiar domains of life such as mating, over someone like Larry King, who has only a high school education, and, incidentally, has had seven wives and five children.

What about the Rest of Us?

The sharp contrast between great success in evolutionarily novel domains and great failures in evolutionarily familiar domains for the very intelligent is not limited to the truly gifted in the Study of Mathematically Precious Youth. General intelligence affects life outcomes in virtually every area and throughout the entire life.[8] From schooling to employment to crime and welfare dependency to civility and citizenship, not only do more intelligent individuals achieve more desirable outcomes, but general intelligence almost always has a linear positive effect on the desirability of the life outcomes. The more intelligent the individuals, the more desirable the life outcomes.

Marriage and parenting are among the very few exceptions to this pattern in a comprehensive review of American life. In fact, "very bright" individuals are the *least* likely to marry of all the cognitive classes. (Recall their definition of cognitive classes at the end of Chapter 3.) Only 67% of these "very bright" white Americans marry before the age of 30, whereas between 72% and 81% of those in other cognitive classes marry before 30.[9] The mean age of first marriage among the "very bright" whites is 25.4, whereas it is 21.3 among the "very dull" individuals and 21.5 among the "dull" individuals. The more intelligent you are, the later you marry.

The pattern is similar in parenting. For example, general intelligence does not confer advantages in giving birth to healthy babies. For example, 5% of white babies born to "very bright" mothers suffer from low birth weight, compared to 1.6% of those born to "bright" mothers and 3.2% of those born to "normal" mothers. Only babies born to "dull" mothers (7.2%) and "very dull" mothers (5.7%) fare worse.[10]

The lack of IQ advantage continues later in the childhood. "Very bright" mothers are more likely to have children who are behind in motor and social development or have the worst behavioral problems. Specifically, 10% of children born to "very bright" white mothers are in the bottom 10% of the motor and social development index, compared to 5% of those born to "bright" mothers and 6% of those born to "normal" mothers.

Similarly, 11% of children born to "very bright" mothers find themselves in the bottom 10% of the behavioral problems index, compared to 6% of those born to "bright" mothers and 10% of those born to "normal" mothers.[11] It is important to note that the problems suffered by children born to "very bright" mothers are not just social and behavioral—for which there might be varying and changing cultural definitions of what constitutes "normal"—but are also physical, such as birth weight and motor development, for which the criteria of normal development are objective and invariant.

Now, since "very bright" white women marry later, and thus give birth to their babies at older ages compared to other mothers, perhaps some of these physical and behavioral problems of their children may be attributable to their older maternal age at birth or the greater possibility of harmful mutations in the sperm of older men.[12] But this is precisely my point. Women with higher intelligence are not using their intelligence to marry early and have healthier children, which are the direct means toward achieving reproductive success. More intelligent individuals are more likely to receive more education and earn more money, both of which are evolutionarily novel, but they are certainly not more likely to achieve the evolutionarily familiar goals of marriage and reproduction. This has led some scholars to muse, "Can mothers be too smart for their own good?"[13]

The Exception that Proves the Rule: Evolutionarily Novel Elements in Mating and Parenting

This is not to argue, however, that intelligent people are not better mates or parents in general today. Intelligent individuals *do* make better mates and parents in some ways in the current (evolutionarily novel) environment. For one thing, more intelligent individuals universally attain more desirable outcomes in all evolutionarily novel domains, such as education, economy, criminal justice, even health and longevity in contemporary society.[14] More educated, wealthier, healthier parents who avoid trouble with the law undoubtedly do make better parents today.

One need go no farther than to recall the news story from many years ago of an illiterate teenage mother whose baby died of dehydration because the mother could not read the instructions for how to make the baby formula and instead fed the dry powder to the baby as is, without first dissolving it in water. Note, however, that this tragedy happened *precisely* because it involved written instructions for making a baby formula—an evolutionarily novel stimulus about an evolutionarily novel product. My contention is that even this mother may have done fine raising her children in the ancestral environment, where childrearing most likely did not require general intelligence, as all the answers necessary for parenting would have been provided by other evolved psychological mechanisms. In the ancestral environment, everyone was illiterate.

Modern means of contraception are another evolutionarily novel element in the evolutionarily familiar domain of mating and parenting. In the ancestral environment, our ancestors probably mated all the time, with pregnancy and lactation (lactational amenorrhea) serving as the only natural means of contraception, besides abstinence. As a result, our ancestors invariably produced a larger number of offspring than we do today, but many of them died in infancy due to infectious diseases, malnutrition, and other natural causes (including predation by humans and other animals).[15] The average number of offspring surviving to sexual maturity in the ancestral environment might not have been much larger than it is today. So while mating and parenting are evolutionarily familiar, voluntary control of fertility through contraception (such as condoms or the pill) is evolutionarily novel. So the Hypothesis would predict that more intelligent individuals are better able to control their fertility voluntarily through artificial means of contraception than less intelligent individuals.

This indeed appears to be the case.[16] Among contemporary Americans in the GSS data, the association between the lifetime number of sex partners and the number of children is *positive* among less intelligent individuals (who are below the median in verbal intelligence), but *negative* among more intelligent individuals (who are above the median in verbal intelligence). The more sex partners less intelligent individuals have, the more children they have, as a natural consequence of greater sexual activity with more partners. In sharp contrast, the more sex partners more intelligent individuals have, the *fewer* children they have. You cannot have fewer children on average by having more sex partners, unless you employ effective contraception.

Figures 4.1 shows the partial association, after controlling for age, race, sex, education, marital status, and religion, between

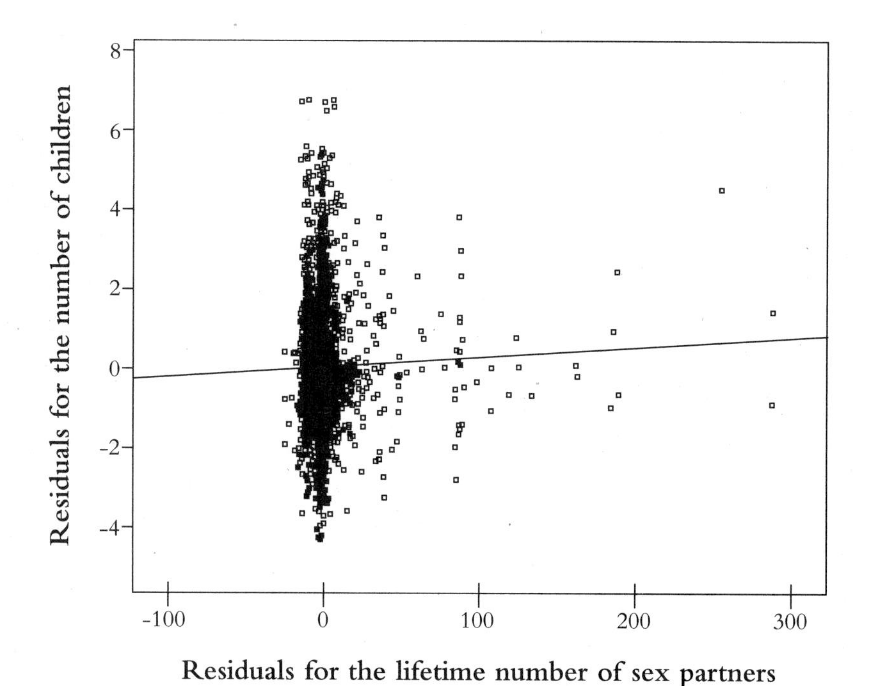

Figure 4.1 Partial association between lifetime number of sex partners and number of children among the less intelligent

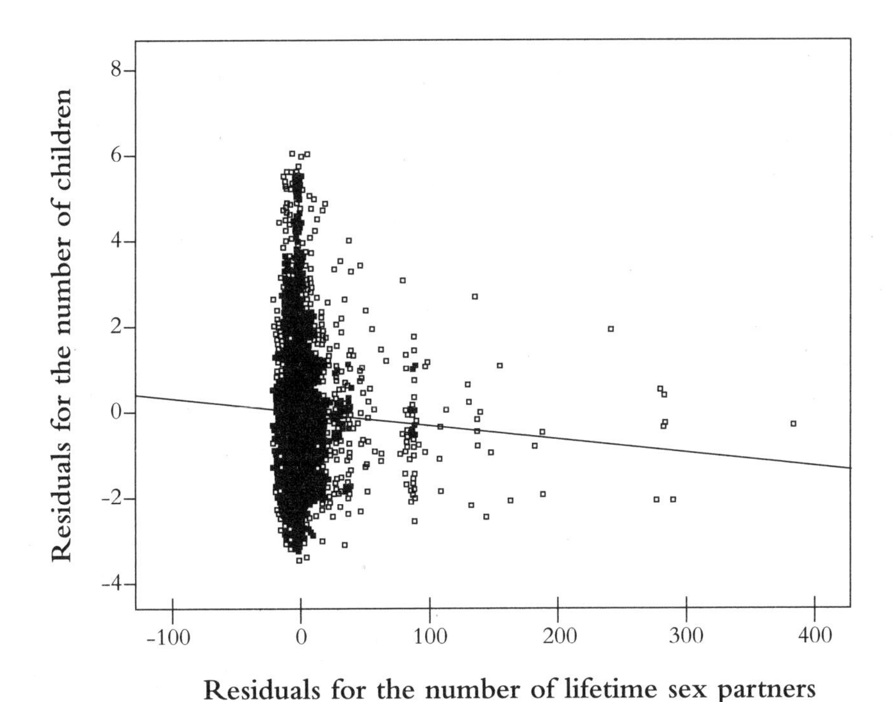

Figure 4.2 Partial association between lifetime number of sex partners and number of children among the more intelligent

the lifetime number of sex partners and the number of children, among individuals who are below the median in verbal intelligence. As you can see, the relationship is positive, as indicated by the regression line with an incline.

Figure 4.2 presents the same relationship among individuals who are above the median in verbal intelligence. Here the relationship between the lifetime number of sex partners and the number of children is negative, as indicated by the regression line with a decline.

The contrast between these two graphs suggests that more intelligent Americans are indeed more efficient in employing (evolutionarily novel) modern means of contraception than their less intelligent counterparts.

Intelligence and Interpersonal Relationships

Interpersonal relationships are an eminently evolutionarily familiar domain of life. Even in the ancestral environment, our ancestors had friends, allies, and enemies that they had to deal with. They also had parents, children, siblings, and other relatives. There is nothing evolutionarily novel about interacting with these categories of people.

In addition, it was very important in the ancestral environment (as it is now) to maintain good relations with these categories of people (except, perhaps, for enemies). Reliable friends and allies are crucial in survival and reproductive success,[17] and investing in kin is a very important means of increasing reproductive success.[18] So the Hypothesis would predict that general intelligence does not have any effect on individuals' ability to maintain interpersonal relationships with these evolutionarily familiar categories of people.

Survey data from the United States support this prediction of the Hypothesis.[19] While more intelligent Americans socialize with their friends significantly more frequently than their less intelligent counterparts, intelligence does not seem to improve interpersonal relationships with other evolutionarily familiar categories of people. In fact, more intelligent individuals socialize with neighbors, siblings, and other relatives significantly *less* frequently than less intelligent individuals. Investing in kin is one of the important means of increasing reproductive success, yet more intelligent American seem less able to do so than less intelligent individuals.

Intelligence and Wayfinding

In the hunter-gatherer life of our ancestors on the African savanna, navigation and wayfinding was an essential skill, on which their very survival depended. After a long hunting or gathering trip, which could sometimes last for days, our ancestors had to find their way home without relying on maps, street signs, artificial landmarks, and the satellite navigation devices. Those who could not find their way home from their trips probably faced certain death. I would therefore expect navigation and wayfinding to be an evolutionarily familiar task, for which there is an evolved psychological mechanism, and the Hypothesis would predict that general intelligence is independent of wayfinding abilities.

A couple of studies support this prediction of the Hypothesis. In a highly ingenious experiment,[20] researchers at York University in Canada took participants on a meandering journey through a wooded area, without any visible landmarks or maps, and asked them, at predetermined locations, to point to the direction of the origin. The participants must then lead the researchers back to the origin. In this study, the participants' wayfinding ability had no correlation at all with their general intelligence, measured by Raven's Progressive Matrices (which, as discussed in Chapter 3, is the best single measure of general intelligence).

Researchers at the University of Arizona replicated the Canadian study in virtual reality.[21] Their participants navigated in computer-generated "rooms" displayed on a computer screen by way of a joystick, and had to find an invisible target placed somewhere in the room on the floor. They got a beep when they (initially unknowingly) "walked" over the invisible target, and then must find it again and again in the same room by navigating to the same location in the room. The researchers' data showed that the participants' general intelligence had no effect on their ability to learn their way around the rooms and return to the invisible target. They concluded that the individual's ability at spatial navigation and general intelligence were largely independent. Given that more intelligent people tend to do virtually everything better than less intelligent people, their distinct lack of advantage in wayfinding is noteworthy.

Intelligence and Exercise

As hunter-gatherers, our ancestors engaged in constant physical activities. Their active, nomadic lifestyle, combined with their limited caloric intake, meant that obesity was probably very rare in the ancestral environment, and most of our ancestors maintained a healthy, active lifestyle by our contemporary standards. It means that regular exercise *for its own sake*, that is, exercising in order to stay healthy and control weight, is probably evolutionarily novel, and the Hypothesis would therefore predict that more intelligent individuals are more likely to engage in regular exercise today than less intelligent individuals.[22]

A couple of recent studies show that this indeed appears to be the case.[23] In one study, the frequency of exercise is significantly positively associated with general intelligence.[24] In another, individuals who can successfully maintain a regular exercise schedule are more intelligent than those who are unsuccessful by more than one full standard deviation (122.50 vs. 106.25). And this is not because intelligence is associated with conscientiousness.[25] It appears that more intelligent individuals are more likely to adopt the evolutionarily novel lifestyle of regular exercise for its own sake.

Of course, more intelligent individuals are more likely to have white-collar desk jobs that are more physically sedentary, whereas less intelligent individuals are more likely to have blue-collar jobs that are physically more active. Thus more intelligent individuals may *need* to exercise more than less intelligent individuals. However, in the first study, education, income, and work status are all statistically controlled, and the participants in the second study are all college students (86.3% of whom are female, who are less likely to have blue-collar jobs). So the effect of intelligence on voluntary physical exercise appears to be genuine, but more research is necessary to arrive at a more firm conclusion.

From the Hypothesis to the Paradox: The Intelligence Paradox on Individual Preferences and Values

I now switch from the discussion of the Savanna-IQ Interaction Hypothesis to the central idea in this book, which I call the *Intelligence Paradox*. The Intelligence Paradox is the application of the Hypothesis to the specific domain of individual preferences and values. It explains how general intelligence affects and influences such preferences and values.

Evolutionarily novel entities and situations that more intelligent individuals are better able to comprehend and deal with may include ideas and lifestyles, which may form the basis of their values and preferences. It would be difficult for individuals to prefer or value something that they cannot truly comprehend. However, comprehension does not equal preference. While not everyone who comprehends certain entities and situations would thereby acquire preferences for them, I assume some would, whereas very few (if any) who do not comprehend them would acquire preferences for them. My assumption is that individuals only prefer or value things that they can truly comprehend. Thus comprehension is a necessary but not sufficient condition for preference.

Applied to the domain of preferences and values, the Hypothesis may therefore suggest what more or less intelligent individuals hold as their preferences and values.[26] Hence I propose the Intelligence Paradox.

> *The Intelligence Paradox: More intelligent individuals are more likely to acquire and espouse evolutionarily novel preferences and values that did not exist in the ancestral environment (and thus our ancestors did not have) than less intelligent individuals. In contrast, general intelligence has*

> *no effect on the acquisition and espousal of evolutionarily familiar preferences and values that existed in the ancestral environment (and thus our ancestors had).*

Recall from Chapter 1 that, by *natural*, I mean "that for which we as a human species are evolutionarily designed," and, by *unnatural*, I mean "that for which we as a human species are not evolutionarily designed." Thus another way to express the Intelligence Paradox is that more intelligent individuals are more likely to acquire and espouse *unnatural* preferences and values which we are not evolutionarily designed to have. Herein lies the essence of the Paradox. More intelligent individuals are more likely to go against their biological design, escape their evolutionary constraints and limitations on their brains, and hence have unnatural and often biologically stupid preferences and values. Yes, more intelligent individuals are more likely to be stupid and do stupid things.

In the remainder of the book, I talk about various applications and manifestations of the Intelligence Paradox with regard to numerous evolutionarily familiar and novel preferences and values. This is where intelligence research meets the problem of values.

Chapter 5

Why Liberals Are More Intelligent than Conservatives

What Is Liberalism?

It is difficult to provide a precise definition of a whole school of political ideology like liberalism or conservatism. To make matters worse, what passes as liberalism or conservatism varies by place and time. The Liberal Democratic Party in the United Kingdom is middle-of-the-road, while the Liberal Democratic Party in Japan is conservative. Most "conservatives" in the UK and Europe are far more liberal than liberal Democrats in the US. The political philosophy which originally emerged as "liberalism" during the Enlightenment is now called "classic liberalism" or "libertarianism," and represents the polar opposite of what is now called "liberalism" in the United States.[1]

In this book, as elsewhere in my work,[2] I adopt the contemporary American definition of liberalism. I provisionally define liberalism (as opposed to conservatism) as *the genuine concern for the welfare of genetically unrelated others and the willingness to contribute larger proportions of private resources for the welfare of such others.* In the modern political and economic context, this willingness usually translates into paying higher proportions of individual incomes in taxes toward the government and its social welfare programs. Liberals prefer higher taxes and income transfers to achieve equality of outcomes. While conservatives believe in equality of opportunities (egalitarianism) and are happy with inequality of outcomes (as long as there is equality of opportunities), liberals believe in equality of outcomes (equality) and prefer any means to achieve it.

Defined as such, liberalism is evolutionarily novel; our ancestors were not liberal in the contemporary American sense. Humans (like all other living species) are designed by evolution to be altruistic toward their genetic kin,[3] their friends and allies with whom they engage in repeated social exchange,[4] and members of their *deme* (a group of intermarrying individuals) or ethnic group.[5] (Yes, it has been mathematically proven that humans are evolutionarily designed to be ethnocentric. The mathematical sociologist Joseph M. Whitmeyer has shown that any individual tendency to benefit those whom one might marry, or those whose children one's children might marry, or those whose grandchildren one's grandchildren might marry—in other words, favoritism toward members of an intermarrying group of people known as the *deme* or, in short, ethnocentrism—will be evolutionarily selected.[6]) But humans are not designed to be altruistic toward an indefinite number of complete strangers whom they are not likely ever to meet or exchange with. This is largely because our ancestors lived in a small band of about 150 genetically related individuals all their lives, and large cities and nations with thousands and millions of people are themselves evolutionarily novel.[7]

But how do we really know that our ancestors were not liberals? In order to make reasonable inferences about what values our ancestors might have held during the course of human evolution, I have relied on two sources. First, I have consulted the 10-volume compendium *The Encyclopedia of World Cultures*,[8] which extensively describes *all* human cultures known to anthropology (more than 1,500) in great detail. Second, I have consulted five different extensive, monograph-length ethnographies of traditional (hunter-gatherer, pastoral, and horticultural) societies around the world.[9] While contemporary hunter-gatherer societies are not exactly the same as our ancestors during the Pleistocene Epoch, they are the best analog that we have available for close examination, and are thus often used for the purpose of making inferences about our ancestral life.

These ethnographic sources make it clear that, while sharing of resources, especially food, with other members of their own tribe is quite common and often expected among hunter-gatherers, and while trade with neighboring tribes may have taken place,[10] there is no evidence that people in contemporary hunter-gatherer bands *freely* share resources *with members of other tribes.* Because all members of a hunter-gatherer tribe are genetic kin (for men) or friends and allies for life (for women),[[11]] sharing of resources among them does not qualify as an expression of liberalism as defined above.

It may therefore be reasonable to infer from these ethnographic accounts that, while sharing of food and other resources with genetic kin may be part of universal human nature, sharing of the same resources with total strangers whom one has never met or is not likely ever to meet is not part of evolved human nature. The Intelligence Paradox would therefore predict that more intelligent individuals are more likely to espouse liberal political ideology than less intelligent individuals.

Are Liberals More Intelligent than Conservatives?

This indeed appears to be the case. Even when statistically controlling for such relevant factors and potential confounds as age, race, education, income, and religion, more intelligent children are more likely to grow up to become more liberal than less intelligent children.[12] Intelligence measured in junior high and high school strongly predicts adult political ideology seven years later. The more intelligent American adolescents are in junior high and high school, the more liberal they become as young adults.

Figure 5.1 shows that young adults in their early 20s who identify themselves as "very conservative" have the average adolescent IQ of 94.82 in junior high and high school, whereas young adults who identify themselves as "very liberal" have the average adolescent IQ of 106.42. And the association between

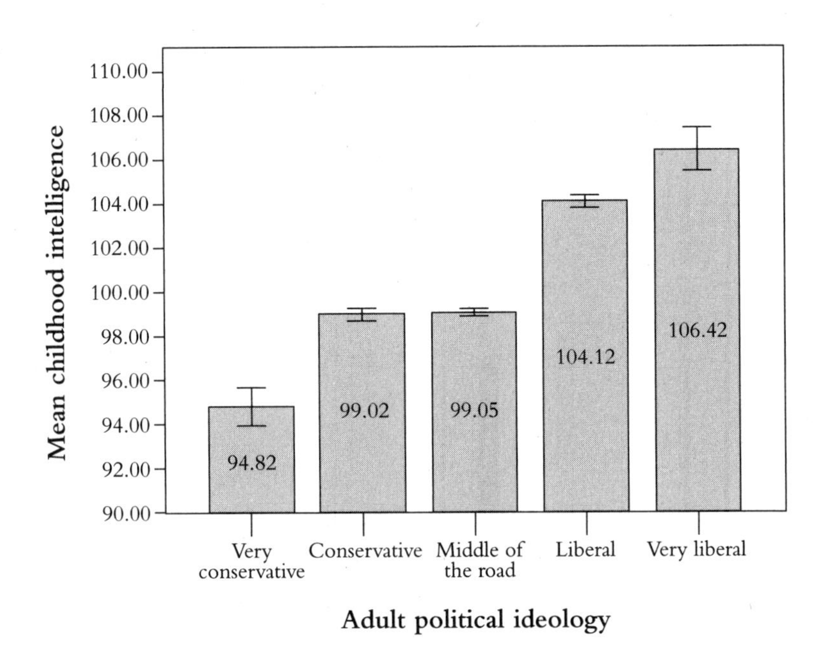

Figure 5.1 Association between childhood intelligence and adult liberalism

adult political ideology and adolescent IQ is monotonic; as one increases, the other steadily increases as well.

As a mean difference between two large categories of individuals, the 11-point difference in childhood IQ between "very liberal" and "very conservative" young American adults is very large and statistically significant. The probability that the results in the above figure can happen by chance, when there is actually no association between intelligence and liberalism, is less than one in 100,000. Even though past studies show that women are more liberal than men,[13] and that blacks are more liberal than whites,[14] the statistical analysis shows that the effect of childhood intelligence on adult political ideology is twice as strong as the effect of sex or race.

The analyses of both Add Health and GSS data confirm the prediction of the Intelligence Paradox that more intelligent individuals are more likely to acquire and espouse the evolutionarily novel value of liberalism, whereas less intelligent individuals are more likely to acquire and espouse the evolutionarily familiar value of conservatism. But the association between intelligence and political ideology is not limited to the United States.

For example, even though there are no "liberals" and "conservatives" in the American sense in the United Kingdom—by the American standard, everybody in the United Kingdom is a socialist and there is very little *substantive* political disagreement among British citizens—a longitudinal study of a large, nationally representative sample of British citizens shows that British children who are more intelligent at ages 5 and 10 are more likely to vote for either the Green Party or the Liberal Democratic Party at age 34.[15] However, because British political parties are not as different from each other on the liberal-conservative dimension as American political parties are, I am not sure what this results means in the context of the Intelligence Paradox.

Incidentally, the empirical finding that liberals are on average more intelligent than conservatives substantiates one of the persistent complaints among conservatives. Conservatives often complain that liberals control the media or show business or academia or some other social institutions. The Intelligence Paradox can explain why conservatives are correct in their complaints. Liberals *do* control the media, or show business, or academia, among other institutions, because, apart from a few areas in life (such as business) where countervailing circumstances may prevail, *liberals control most institutions.* They control the institutions, not because they are liberals, but because they are on average more intelligent than conservatives. They are therefore more likely to attain the highest status in any areas of evolutionarily novel modern life.

Of course, as with any other broad empirical generalizations, there *are* exceptions. Liberals do not control every single organization in every single area of life. However, the balance is far from equal, and the overall bias is pretty obvious. For example, in the area of the mass media, AM talk radio is predominantly conservative; however, it is only one of many mass media channels (newspapers, magazines, TV, radio, movies, the internet, etc.), and all other mass media exhibit a strong liberal bias. In fact, as I elaborate later in Chapter 10, while *AM* talk radio is conservative, *FM* talk radio (mostly, NPR) is very liberal. Among cable TV news channels, Fox News is relatively conservative, but it is only one of many cable TV news channels, all of which are entirely liberal. So it is safe to conclude that liberals control *most* organizations in *most* areas of life, even though the American population in general is mostly conservative.

If Liberals Are More Intelligent than Conservatives, Why Are Liberals So Stupid?

While it is consistent with the prediction of the Intelligence Paradox, the conclusion in the previous section that liberals are on average more intelligent than conservatives may not resonate with most people's daily observations and experiences. If they are more intelligent, why are liberals—especially those in Hollywood and academia—so much more likely than conservatives to say and do stupid things and hold incredulous beliefs and ideas that stretch credibility?

Bruce G. Charlton, Professor of Theoretical Medicine at the University of Buckingham and former Editor in Chief of *Medical Hypotheses*, may have an explanation. In his editorial in the December 2009 issue of *Medical Hypotheses*,[16] Charlton suggests that liberals and other intelligent people may be "clever sillies," who incorrectly apply abstract logical reasoning to social and interpersonal domains. As I explain in Chapter 3, general intelligence—the ability to think and reason—likely evolved as a domain-specific evolved psychological mechanism to solve evolutionarily novel problems, whereas, for all evolutionarily familiar problems, there are other dedicated psychological adaptations. Everyone—intelligent or not—is evolutionarily designed to have the ability to solve such evolutionarily familiar problems in the social and interpersonal domains as mating, parenting, social exchange, and personal relationships with the other evolved psychological mechanisms.

Charlton suggests that the totality of all the other evolved psychological mechanisms (all of human nature except for general intelligence) represents what we normally call "common sense." Everyone has common sense. Intelligent people, however, have a tendency to overapply their analytical and logical reasoning abilities derived from their general intelligence incorrectly to such evolutionarily familiar domains and, as a result, get things wrong. In other words, liberals and other intelligent people lack common sense because their general intelligence overrides it. They *think* in situations where they are supposed to *feel*. In evolutionarily familiar domains such as interpersonal relationships, feeling usually leads to correct solutions whereas thinking does not.

I personally dislike Charlton's term "clever sillies"—I don't like the British usage of both words, "clever" and "silly." But otherwise I completely agree with his analysis substantively. As Charlton points out, common sense is eminently evolutionarily familiar. Our ancestors could not have survived a single day in their hostile environment full of predators and enemies if they did not possess functional common sense. That's why it has become an integral part of evolved human nature in the form of evolved psychological mechanisms in the social and interpersonal domains.

As I explain in my last book,[17] despite all the surface and superficial differences, all human cultures are essentially the same in broad and abstract terms; there is only one human culture. And part of the common human culture is "common sense" about how to behave and how to treat each other. So not only do individuals from different ethnic, cultural, political, and class backgrounds in the same society share common sense, but so do all peoples of the world. Notice that common sense only pertains to evolutionarily familiar and relevant aspects of social life, not evolutionarily novel aspects. There is no common sense about how to boot up a Macintosh computer or how to fly an airplane, although there is common sense about how to behave in a computer lab or in a crowded airplane, which is the same as how to behave in a crowded cave. Common sense is thus evolutionarily familiar.

Because common sense is evolutionarily familiar and thus natural, the Intelligence Paradox would predict that more intelligent people may be less likely to resort to it. They may be more likely to resort to evolutionarily novel, non-commonsensical, stupid ideas to solve problems in the evolutionarily familiar domains. If this is not paradoxical, worthy of the name "the Intelligence Paradox," I don't know what is.

This, incidentally, is the reason I never use words like "smart" and "clever" as synonyms for "intelligent." Similarly, I never use words like "dumb" and "stupid" as synonyms for "unintelligent."

"Intelligent" has a specific scientific meaning—possessing higher levels of general intelligence measured by a series of cognitive tests or heavily *g*-loaded tests like Raven's Progressive Matrices, as I explain in Chapter 3. In sharp contrast, "smart" and "stupid" have more to do with common sense than intelligence. From my perspective, more intelligent people like liberals are more likely to be "stupid" (lacking common sense), whereas less intelligent people like conservatives are more likely to be "smart" (possessing functional common sense). Yes, more intelligent people are stupider, and less intelligent people are smarter. If this is not paradoxical, I don't know what is.

Matt Stone and Trey Parker—the co-creators of *South Park*—get it perfectly. In the episode "Go God Go XII," the Wise One (the elderly leader of atheist otters—Don't ask: You have to see it) says, with reference to Richard Dawkins:

> Perhaps the Great Dawkins wasn't so wise. Oh, he was intelligent, but some of the most intelligent otters that I've ever met were completely lacking in common sense.

Charlton's concept of "clever sillies," and the Intelligence Paradox, can explain why general intelligence and the capacity for common sense may be negatively associated across individuals, and why people like Richard Dawkins are simultaneously very intelligent and very stupid, lacking in common sense, *precisely because* they are very intelligent. As the pioneer evolutionary psychologist Gordon G. Gallup Jr. puts it very eloquently, in science, common sense is often common nonsense.[18]

Higher Intelligence as a Peacock's Tail?

There may be other reasons why intelligent people like liberals tend to espouse stupid ideas. The Norwegian-Australian journalist Mads Andersen, in personal communication, suggests to me another explanation for why liberals are stupider than conservatives. Andersen has a couple of great suggestions, both of which utilize the *handicap principle*, first proposed by the Israeli biologist Amotz Zahavi.[19]

A prime example of a handicap is the peacock's tail. The long, elaborate, and ornate tail of a peacock does not have any adaptive value; it does not serve any tangible, useful purpose that would aid the survival of the peacock. In fact, it only harms its survival chances. Peacocks with longer, more elaborate trains are easier for predators to catch and kill than fellow peacocks with shorter and simpler trains. It is also biologically more expensive to maintain elaborate trains with symmetrical eyes. So they only have costs and no benefits.

But that, according to Zahavi, is precisely the point. Peacocks are advertising to peahens, "Look, I am so genetically fit and I can run so fast that I can still evade the predators with this huge thing hanging from my ass! Them other guys ain't so fit and the only reason they can evade predators is because their trains are shorter. They wouldn't be able to evade the predators if their tails were as long as mine! Now whose genes would you like your offspring to carry?"

And peahens indeed do prefer to mate with peacocks with longer, more elaborate, and more symmetrical tails that are biologically very expensive to maintain and costly for their survival chances. Peahens prefer such peacocks as mates so that their male offspring will also sport long, elaborate tails that attract females of their generation.

The same idea is captured by the expression "fighting with one arm tied behind my back." Any fighter who can win a fight with one arm tied behind his back would naturally have to be stronger and more genetically fit than anyone who needs both hands to fight. Zahavi and other biologists suggest that many seemingly useless traits like peacocks' long tails may have evolved as a handicap, an honest signal of one's genetic fitness to potential mates. They are therefore sexually selected (they increase the carrier's reproductive success), even though they are not naturally selected (they do not increase the carrier's survival chances).

Andersen's ideas capitalize on the Zahavian handicap principle. First, he suggests that more intelligent individuals tend to espouse absurdly complex ideas as an honest signal of their higher intelligence. Because common sense is evolutionarily familiar, and all humans are equipped with common sense, it is by definition the simplest and easiest solution available to them. More intelligent people reject the "simplistic" solution offered by common sense, even though it is usually the correct solution, and instead adopt unnecessarily complex ideas simply because their intelligence allows them to entertain such complex ideas, even when they may be untrue or unuseful in solving the problem at hand.

Many observers have noted that this is indeed already happening in academia.[20] In fields like literary criticism that lack external objective criteria for evaluating ideas (in contrast to natural sciences whose theories must be evaluated against nature), or in pseudoscientific fields like sociology where nobody can agree on what the truth is and political ideology trumps empirical evidence, academics are increasingly rewarded for proposing complex and absurd ideas like reader response theory or social constructionism. Andersen suggests that these academics may be (unconsciously) saying, "Look, I have such an excess of intelligence that I don't have to go for the obvious and simple (albeit true) answers provided by common sense. I can come up with absurdly complex ideas because my higher intelligence allows me to!"

Second, Andersen points out that many political liberals, especially in Hollywood and academia, are themselves well off and do not individually and directly benefit from the liberal policies of greater welfare states. Once again, these liberals may be giving an honest signal that they have accumulated and are still able to accumulate so many resources that they can afford to pay higher taxes and allow the public funds to benefit other people. If they are not able to accumulate resources themselves, they would not be able to afford paying higher taxes to fund welfare programs that do not directly benefit them. In essence, they are (unconsciously) saying, "Look, I'm so wealthy that I can afford to waste my money on other people who are not related to me!"

I believe Andersen may be right in both of his suggestions. But I don't think his explanations necessarily contradict the ones offered by Charlton and myself earlier. Instead, they may be additional reasons why intelligent people are more likely to be liberals and espouse stupid ideas as honest signals of their genetic fitness and higher intellectual capacity. One does not have to be wrong for the other to be right; they may both be correct and provide partial explanations.

IQ and the Values of Nations

The Intelligence Paradox about the effect of general intelligence on individual preferences and values may also have implications for national differences in their characters, institutions, and laws. Just like individuals, nations also vary in their collective preferences and values. Individual preferences and values of millions of people can aggregate to shape national institutions and laws at the societal level.

If more intelligent individuals are more likely to be liberal, as the data both in the United States and the United Kingdom seem to suggest, then it logically follows that, at the societal level, populations with higher average intelligence are more likely to be liberal. Data indeed do confirm this macrolevel implication of the Intelligence Paradox.

Even after statistically controlling for such relevant factors as economic development, education, history of communism, geographic location, and the size of the government, societies with higher average intelligence are more liberal.[21] The average intelligence in society increases the highest marginal tax rate (as an expression of people's willingness to contribute their private resources for the welfare of genetically unrelated others) and, partly as a result, decreases income inequality. The more intelligent the population, the more they pay in income taxes and the more egalitarian their income distribution.

In fact, the average intelligence of the population is the strongest determinant of the highest marginal tax rate and the level of income inequality in the society. Each IQ point in average intelligence increases the highest marginal income tax rate by more than half a point; in societies with higher average intelligence by 10 IQ points, individuals pay more than 5% more of their individual incomes in taxes.

It appears that the Intelligence Paradox can not only explain individual differences in their preferences and values, but also national differences in their characters, institutions, and laws, in other words, the values of nations.

Chapter 6

Why Atheists Are More Intelligent than the Religious

Where Does Religion Come From?

While religion is a cultural universal—humans in all known societies practice some religion[1]—recent evolutionary psychological theories suggest that religiosity—belief in higher powers—may not be an adaptation in itself.[2] Religiosity may instead be a byproduct of other evolved psychological mechanisms, variously known as "animistic bias"[3] or "the agency-detector mechanism."[4] Now, what in God's name does that mean?

Imagine you are our ancestor living on the African savanna 100,000 years ago, and you encounter some *ambiguous situation.* For example, you heard some rustling noises nearby at night. Or you were walking in the forest, and a large fruit falling from a tree branch hit you on the head and hurt you. Now what's going on?

Given that the situation is inherently ambiguous, you can either attribute the phenomenon to impersonal, inanimate, and unintentional forces (for example, wind blowing gently to make the rustling noises among the bushes and leaves, or a mature fruit falling by the force of gravity and hitting you on the head purely by coincidence) or attribute it to personal, animate, and intentional forces (for example, a predator hiding in the dark and getting ready to attack you, or an enemy hiding in the tree branches and throwing fruits at your head to hurt you). The question is, which is it?

As you can see in the 2 × 2 table in Figure 6.1, there are four possible outcomes. In the two diagonal cases, you have made the correct inference. You inferred that the cause of the ambiguous situation was personal, animate, intentional, and it was; or you inferred that the cause of it was impersonal, inanimate, unintentional, and it was. There are no negative consequences if you made the correct inference.

		True state of nature	
		Personal, animate, intentional	**Impersonal, inanimate, unintentional**
Inference	**Personal, animate, intentional**	Correct inference	False positive (Type I) error Consequence = paranoia
	Impersonal, inanimate, unintentional	False negative (Type II) error Consequence = potential death	Correct inference

Figure 6.1 Error management theory applied to religiosity

Given the insufficient information you have, however, you cannot always make the correct inference. Sometimes you make mistakes in your judgment. In the two off-diagonal cases, you have made incorrect inferences. If you inferred that the cause was personal, animate, and intentional, whereas its true cause was impersonal, inanimate, and unintentional, you have made what the statisticians call the "Type I" error of false positive. You thought the danger was there, when it wasn't. In contrast, if you inferred that the cause was impersonal, inanimate, and unintentional, whereas its true cause was personal, animate, and intentional, you have made what the statisticians call the "Type II" error of false negative. You didn't think there was danger, when there was.

All errors in inference have negative consequences, but these two types of errors—Type I error of false positive and Type II error of false negative—have very different negative consequences. The consequence of Type I error is that you become paranoid. You are always looking around and behind your back for predators and enemies that don't exist. The consequence of Type II error is that you are dead, being killed by a predator or an enemy when you least expect them. Obviously, it's better to be paranoid than dead, so evolution should have designed an inference system that *overinfers* personal, animate, and intentional forces even when none exist. Evolution should build an inference system that minimizes the chances of making Type II errors.

Here's the catch. An inference system cannot simultaneously decrease the chances of making Type I error and Type II error. Any inference system that decreases the probability of making Type I error must necessarily increase the probability of making Type II error. Conversely, any inference system that decreases the probability of making Type II error must necessarily increase the probability of making Type I error. So if the human mind has been selected to minimize the probability of making Type II errors, so that they could never be caught off guard and attacked by predators and enemies that they assumed didn't exist, then the human mind must necessarily make a large number of Type I errors.

You cannot simultaneously be paranoid and oblivious (or relaxed). The more paranoid you are, then, necessarily, the less oblivious (or relaxed) you are. The more oblivious (or relaxed) you are, then, necessarily, the less paranoid you are. In the face of a potentially dangerous yet ambiguous situation, the human mind is designed to be more paranoid and less oblivious.

Think of a smoke detector, which is designed, not by evolution by natural and sexual selection, but by engineers.[5] Just like the human mind's inference system, smoke detectors can make errors of inference. It can sound the alarm, "thinking" there is fire, when there isn't (Type I error of false positive), or it can remain silent, "thinking" there is no fire, when there is (Type II error of false negative). The consequence of Type I error is that you are woken up in the middle of the night by the fire alarm, when there is no fire. The consequence of Type II error is that you sleep through the fire and are potentially burned to death.

		True state of nature	
		Fire	**No fire**
Inference	**Fire!**	Correct inference	False positive (Type I) error Consequence = woken up in the middle of the night
	No fire	False negative (Type II) error Consequence = potential death	Correct inference

Figure 6.2 The smoke detector principle

As annoying as it is to be repeatedly woken up by false alarms at three o'clock in the morning, the annoyance, even repeated annoyance, is nothing compared to what could happen if the smoke alarm makes one Type II error of not sounding the alarm when there is fire. So the smoke detector designers and engineers deliberately design smoke detectors to make lots of Type I errors of sounding alarm when there is no fire, in order to make sure that it would never ever make a single fatal Type II error of remaining silent when there is fire. Smoke detectors are therefore designed to be extremely sensitive for any potential smoke or fire. Just like the human mind, smoke detectors are designed to be "paranoid." This is known as the "smoke detector principle."[6]

Recent evolutionary psychological theories therefore suggest that the human inference system may have been designed by evolution to operate like a smoke detector. It may be designed to make as few Type II errors as possible, and, as a necessary and unavoidable consequence, to make many Type I errors. These theories suggest that the evolutionary origins of religious beliefs in supernatural forces may have come from such an innate cognitive bias to commit Type I errors rather than Type II errors, and thus to overinfer personal, animate, intentional forces behind otherwise perfectly natural phenomena. This tendency underlies what some researchers call the "animistic bias" or the "agency-detector mechanism." These tendencies happen because evolution employs the same "smoke detector principle" that engineers use.

You see a bush on fire. It could have been caused by an impersonal, inanimate, and unintentional force (lightning striking the bush and setting it on fire). Or it could have been caused by a personal, animate, and intentional force (God trying to communicate with you). The "animistic bias" or "agency-detector mechanism" predisposes you to opt for the latter explanation rather than the former. It predisposes you to see the hands of

God (an animistic and intentional agent) at work behind natural, physical phenomena whose exact causes are unknown.

In this view, religiosity—the human capacity for belief in supernatural beings—is not an evolved tendency per se; after all, religion itself is not adaptive. It is instead a *by-product* of the animistic bias or the agency-detector mechanism, the tendency to be paranoid, which *is* adaptive because it can save your life. Humans did not evolve to be religious; they evolved to be paranoid. And humans are religious because they are paranoid.

Some readers may recognize this argument as a variant of "Pascal's wager." The 17th-century French philosopher Blaise Pascal (1623–1662) argued that given that one cannot know for sure if God exists, it is nonetheless rational to believe in God. If one does not believe in God when He indeed exists (Type II error of false negative), one must spend eternity in hell and damnation. In contrast, if one believes in God when he actually does not exist (Type I error of false positive), one only wastes a minimal amount of time and effort spent on religious services. The cost of committing Type II error is much greater than the cost of committing Type I error. Hence one should rationally believe in God.

Is It Natural to Believe in God?

So recent evolutionary psychological theories suggest that the evolutionary origin of religious beliefs in supernatural forces may stem from such an innate evolutionary bias to commit Type I errors rather than Type II errors. If these theories are correct, then it means that religion and religiosity have an evolutionary origin. It is evolutionarily familiar and natural to believe in God, and evolutionarily novel not to be religious.

Once again, in order to make reasonable inferences about the religious beliefs of our ancestors during the course of human evolution, I consult the same primary ethnographic sources on which I rely in Chapter 5 in making inferences about their liberalism. Out of more than 1,500 distinct cultures throughout the world described in *The Encyclopedia of World Cultures*,[7] only 19 contain any reference to atheism Not only do all these 19 cultures exist far outside of our ancestral home in sub-Saharan Africa, but all 19 without an exception are former Communist societies (Abkhazians in Georgia, Ajarians in Georgia, Albanians, Bulgarians, Chuvash in Russia, Czechs, Germans in Russia [but not in Germany], Gypsies in Russia, Itelmen in Russia, Kalmyks in Russia, Karakalpaks in Russia, Koreans in Russia [but not in Korea], Latvians, Nganasan in Russia, Nivkh in Russia, Poles, Turkmens, Ukrainian peasants, Vietnamese). All Communist states are officially atheist and impose atheism on their citizens. There are no non-former-Communist cultures described in *The Encyclopedia* as containing any significant segment of atheists.

Nor is there any reference to any individuals who do not subscribe to the local religion in any of the monograph-length ethnographies that I have consulted.[8] Once again, contemporary hunter-gatherers are not exactly the same as our ancestors during evolutionary history. But I do believe it is quite telling that, out of all the human cultures known to anthropology (more than 1,500 cultures), the *only* cultures that contain a substantial number of atheists (only 19 out of 1,500+) are former communist societies. It may therefore be reasonable to conclude that atheism may not be part of the universal human nature, and widespread practice of atheism may have been a recent product of Communism in the 20th century. The Intelligence Paradox would therefore suggest that more intelligent individuals are more likely to be atheist than less intelligent individuals.

Consistent with the prediction of the Intelligence Paradox, analyses of Add Health and GSS data show that more intelligent children are more likely to grow up to be atheists.[9] Even after statistically controlling for such relevant factors as age, sex, race, education, income, and religion, more intelligent individuals are more likely to be atheistic than less intelligent individuals.

Even though intelligence and education are highly positively correlated (because, as I explain in Chapter 3, more intelligent individuals on average receive more education), intelligence and education have opposite effects on religiosity. More intelligent individuals are *less* religious, while more educated individuals are *more* religious (net of intelligence and all the other variables mentioned above). So it is decidedly *not* that more intelligent individuals are less religious because they are more educated and education reduces religiosity. First, education is statistically held constant (its effect is removed) in assessing the effect of intelligence on religiosity. Second, contrary to what you might expect, more educated people are *more* religious, not less.

As you can see in Figure 6.3, young adults (in their early 20s) who identify themselves as "very religious" have the average childhood IQ of 97.14 in junior high and high school. In contrast, young adults who identify themselves as "not at all religious" have the average childhood IQ of 103.09. Once again, because these are averages from a sample of tens of thousands of Americans, the difference of 6 IQ points separating the two extreme categories is very large and statistically significant. The probability that the results in Figure 6.3 can happen by chance, when there is actually no association between intelligence and religiosity, is less than one in 100,000.

Even though past studies have shown that women are much more religious than men,[10] the analysis shows that the effect of childhood intelligence on adult religiosity is twice as strong as the effect of sex. It is remarkable that childhood intelligence is a significant determinant of adult religiosity even when religion itself (whether the respondent is Catholic, Protestant, Jewish, or other) is statistically controlled for (with "no religion" as the reference category).

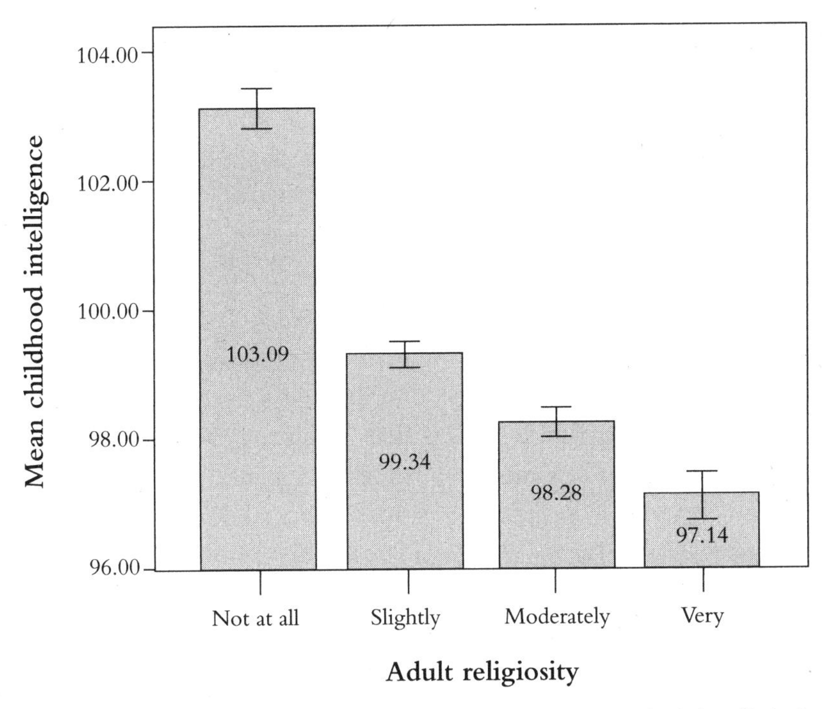

Figure 6.3 Association between childhood intelligence and adult religiosity

Societal Implications

Once again, if more intelligent individuals are more likely to be atheistic, then it follows that, at the societal level, the average intelligence of a population should decrease its average level of religiosity. The more intelligent the population, the less religious (and more atheistic) it should be. This indeed appears to be the case.[11]

Across nations, even after statistically controlling for such relevant factors and potential confounds as economic development, education, history of communism, and geographic location, the average intelligence of a population significantly and strongly reduces its average level of religiosity. Average intelligence in society decreases the proportion of the population who believe in God, how important God is to the people, and the proportion of the population who consider themselves to be religious. The more intelligent the population, the less religious they are on average.

In fact, just as with the highest marginal tax rate and income inequality discussed in Chapter 5, the average intelligence of the population is the *strongest* determinant of its level of religiosity. Each IQ point in average intelligence, for example, decreases the percentage of the population who believe in God by 1.2%, and the proportion of people who consider themselves to be religious by 1.8%. Average intelligence singlehandedly explains 70% of the variance in how important God is in different nations.

It appears, once again, that the Intelligence Paradox not only explains individual preferences and values, such as religiosity, at the micro level of individuals, but can also explain national differences in values, preferences, and character at the macro level of societies.

Chapter 7

Why More Intelligent Men (but Not More Intelligent Women) Value Sexual Exclusivity

Humans Are Naturally Polygynous, Not Monogamous

Because there is still a lot of confusion over terminology, I feel I must repeat what I said in my earlier book and first clarify different institutions of marriage.[1] *Monogamy* is the marriage of one man to one woman, *polygyny* is the marriage of one man to more than one woman, and *polyandry* is the marriage of one woman to more than one man. *Polygamy*, even though it is often used in common discourse as a synonym for polygyny, refers to both polygyny and polyandry. It is therefore ambiguous as to what polygamy refers to, and the word should be avoided in any scientific discussion, unless it refers specifically to both polygyny and polyandry simultaneously. The reason people use "polygyny" and "polygamy" interchangeably is because, for reasons I explain in the earlier book,[2] there are very few polyandrous societies in the world (polyandry contains the seed of its own extinction), and virtually all polygamous societies are indeed polygynous.

Throughout most of human evolutionary history, our ancestors were mildly polygynous, not monogamous. Now how do we know this? Marriage institutions, unlike skeletons, do not leave fossil records. So how do we know what kind of marriage institution (polygyny vs. monogamy) our ancestors practiced? From their skeletons; that's how we know.

How polygynous members of a species are in general correlates with the extent of sexual dimorphism in size (the average size difference between the male and the female). The more sexually dimorphic the species (where the males are bigger than the females), the more polygynous the species.[3] This is *either* because males of polygynous species become larger in order to compete with other males and monopolize females,[4] *or* because females of polygynous species become smaller in order to mature early and start mating.[5] Sexual selection can also create sexual dimorphism in size, if women prefer taller men as mates and/or if men prefer shorter women as mates.

I personally believe that polygyny is key to sexual dimorphism in size among humans.[6] I believe men and women could potentially be the same size in a perfectly monogamous society. But all human societies are invariably polygynous to various degrees.[7] In fact, women's (but not men's) average height in society is partly determined by its degree of polygyny. The more polygynous the society, the shorter women are on average, while men's average height is unaffected.[8]

At any rate, what is indisputable is the positive association between the degree of polygyny and the degree of sexual dimorphism in size, both across species and across human societies. Thus strictly monogamous gibbons are sexually monomorphic (males and females are about the same size), whereas highly polygynous gorillas are equally highly sexually dimorphic in size.

Southern elephant seals (*Mirounga leonina*) represent the extreme of this association. Male southern elephant seals on average maintain about 50 females in their harems. In other words, only 2% of male southern elephant seals get to reproduce each breeding season, and the other 98% end up being complete reproductive losers. Theirs is an extremely polygynous breeding system. As a result, male southern elephant seals are nearly eight times as large by weight as female elephant seals.[9] In fact, many female southern elephant seals are often crushed to death under the weight of the male during copulation.

Fortunately, the average human male is only 17% heavier than the average human female.[10] So, on this scale, humans are *mildly* polygynous, not as polygynous as gorillas (let alone southern elephant seals), but not strictly monogamous like gibbons either.

Consistent with this comparative evidence, a comprehensive survey of traditional societies shows that an overwhelming majority (83.39%) of them practice polygyny, with only 16.14% practicing monogamy and .47% practicing polyandry.[11] The fact that polygyny is so widespread in such societies, combined with the comparative evidence discussed above, strongly suggest that our ancestors might have practiced polygyny throughout most of human evolutionary history.

Of course, polygynous marriage in any society is mathematically limited to a minority of men.[12] Given a roughly 50–50 sex ratio, the highest proportion of men in polygynous marriage in any society is 50%. If half the men each take two wives, the other half must remain wifeless. If some men take more than two wives, more men must remain wifeless and the proportion of polygynous men will be even smaller. So the proportion of polygynous men in any society must always be lower than 50%. Most men in polygynous societies either have only one wife or no wife at all.

However, at least some men throughout evolutionary history were polygynous, and we are disproportionately descended from polygynous men with a large number of wives, because such men had more children than monogamous or wifeless men. Nor does the human evolutionary history of mild polygyny mean that women have always remained faithful to their legitimate husband. There is clear anatomical evidence in men's genitals to suggest that women have always been mildly promiscuous over human evolutionary history.[13]

As you can see in Figure 7.1, under polygyny, one man is married to several women, so a woman in a polygynous marriage still (legitimately) mates with only one man, just as a woman in a monogamous marriage does. So a woman in a polygynous marriage and a woman in a monogamous marriage are both (supposed to be) sexually exclusive to one man. In sharp contrast, a man in a polygynous marriage concurrently mates with several women, quite unlike a man in a monogamous marriage, who mates with only one woman. So throughout human evolutionary history, men have mated with several women concurrently while women have (legitimately) mated with only one man.

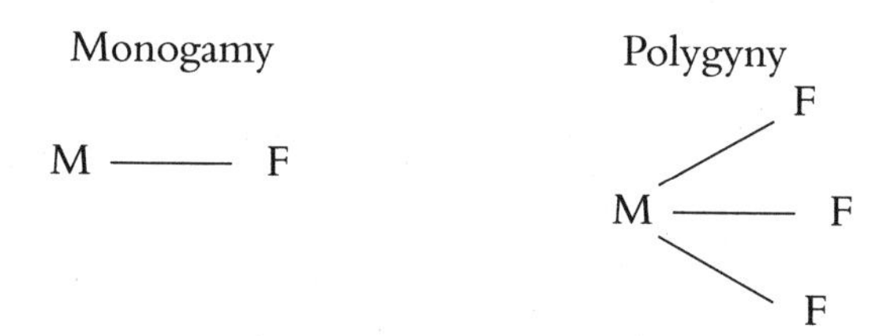

Figure 7.1 Marriage institutions: Monogamy vs. polygyny

Sexual exclusivity prescribed under socially imposed monogamy today is therefore evolutionarily novel for men, but not for women. The Intelligence Paradox would therefore predict that more intelligent men may value sexual exclusivity—having only one sexual partner in a committed relationship—more than less intelligent men, but intelligence may not affect women's likelihood of espousing value of sexual exclusivity.

Consistent with this prediction of the Intelligence Paradox, an analysis of the Add Health data shows that more intelligent boys are more likely to grow up to value sexual exclusivity in early adulthood than less intelligent men. Even after statistically controlling for the effects of such relevant factors and potential confounds as age, race, education, income, religion, and the number of times the respondent has been married, the more intelligent they are in junior high and high school, the more they value sexual exclusivity in a relationship seven years later. In contrast, net of the same control variables, childhood IQ does not affect women's value on sexual exclusivity in early adulthood (in their 20s). More intelligent girls do not grow up to value sexual exclusivity in a relationship more than less intelligent girls. The effect of childhood intelligence on the value of sexual exclusivity is more than four times as strong among men as among women. Among women, the association between childhood intelligence and adult value on sexual exclusivity is not statistically significant at all.

Are More Intelligent Men More Likely to Be Faithful?

Some tabloid newspapers have sensationally reported the above findings with such salacious headlines as "Cheat-on-Wives Men 'Less Intelligent'"[14] (*Metro*), "Smart Men Less Likely than Dumb

Ones to Cheat on Lovers: Study"[15] (*New York Daily News*), and "Intelligent Men 'Less Likely to Cheat'"[16] (*Daily Telegraph*). Apart from the sensationalism, and the confusion between "intelligent" and "smart," which I address in Chapter 5, are these headlines correct? Do the data presented above suggest that more intelligent men are less likely to be sexually unfaithful to their wives and girlfriends?

Probably not. But before I can explain why not, and why more intelligent men are actually *more* likely to cheat, I need to make a small detour and discuss a very important concept in evolutionary biology called *female choice.*

Female Choice

Among all mammalian species (including humans) in which the female makes greater parental investment in children than the male does, sex and mating are a female choice, not a male choice.[17] It happens whenever and with whomever the female wants, not whenever and with whomever the male wants. And humans are no exception.

A couple of studies brilliantly highlight the operation of female choice among humans. In a classic 1989 study, two social psychologists, Russell D. Clark and Elaine Hatfield, hired a young attractive confederate of each sex to approach college students of the opposite sex on campus.[18] So a young attractive female confederate would approach a male student, or a young attractive male confederate would approach a female student, and say, "I have been noticing you around campus. I find you to be very attractive." Then the confederate asked one of three questions: "Would you go out with me tonight?" "Would you come over to my apartment tonight?" and "Would you go to bed with me tonight?" The confederate would then simply record the subject's response (yes or no). They conducted the experiment twice, once in 1978 and again in 1982.

Here's what Clark and Hatfield found.

1978 Percentage Saying "Yes"

	Type of request		
Sex of requestor	**Date**	**Apartment**	**Sex**
Male	56%	6%	0%
Female	50%	69%	75%

1982 Percentage Saying "Yes"

	Type of request		
Sex of requestor	**Date**	**Apartment**	**Sex**
Male	50%	0%	0%
Female	50%	69%	69%

There are two interesting findings here, although they are wholly unsurprising to anyone with common sense. First, both in 1978 and 1982, *absolutely none* of the dozens of women who were approached by a handsome strange man agreed to have sex with him. Second, an overwhelming majority of men (75% in 1978 and 69% in 1982) agreed to have sex with a beautiful strange woman whom they had never met before. Notice that a much smaller proportion of men (exactly 50% in both 1978 and 1982) would go out on a date with her. In other words, many men who would not go out on a date with the woman would nonetheless have sex with her! More women than men in 1978 and exactly as many women as men in 1982 were willing to go out on a date with the stranger, but none of the women would sleep with him.

This classic study was recently replicated in Denmark in 2009, even though the sex difference in the proportion saying "yes" (2% vs. 38%) is not as stark in the Danish study as it was in the original American study (0% vs. 75%).[19] Interestingly, while *none* of the Danish women who were not currently in relationships said yes (just like their American counterparts), 4% of the women who were currently in relationships did. Given that the male confederates who approached them were physically attractive, this finding is actually consistent with the evolutionary psychological prediction from the Good Genes Sexual Selection Theory.[20] It proposes that women seek to marry resourceful men of high status who would invest in their children ("dads") while at the same time being impregnated by handsome men of high genetic quality ("cads"), and then to pass off the resultant children as their long-term mates'. The women cannot cuckold their mates unless they are already in a relationship, so it makes sense from this perspective that more women who are in relationships agreed to sleep with a handsome stranger.

Many of the men who said "No" in the original Clark and Hatfield study actually *apologized* to the female confederate, saying that they could not sleep with her because they were married or have a steady girlfriend, implying that they would have slept with her if they weren't married or didn't have a steady girlfriend. The Danish study indeed shows that men who were not currently in relationships were much more likely to say yes than men who were (59% vs. 18%). In contrast, many of the women were *angry* when the male confederate asked them if they would sleep with him.

The Clark and Hatfield study, like nothing else, demonstrates the power of female choice, which is why it has become a classic. It shows that a reasonably attractive young woman can approach any man and have sex with him. A reasonably attractive young man cannot with a woman. The woman decides when and with whom sex takes place; the man doesn't.

Another study shows the continuing power of female choice. In an article published in the September 2008 issue of *Obstetrics & Gynecology*,[21] Bliss Kaneshiro of the University of Hawaii and colleagues studied the effect of women's body mass index (BMI) on their sexual behavior. Their sample contained 3,600 women of "normal" body weight (BMI < 25), 1,643 "overweight" women

(25 < BMI < 30), and 1,447 "obese" women (BMI > 30) between the ages of 15 and 44.

Their statistical analysis shows that there is no significant difference between normal-weight women, on the one hand, and overweight and obese women, on the other, on their sexual orientation, age at first intercourse, frequency of heterosexual intercourse, and the number of either lifetime or current male sexual partners. It means that, contrary to what one might expect, overweight and obese women are *not* having sex later, less frequently, or with fewer partners than normal-weight women. There *is* a significant difference, however, on whether they have ever had sexual intercourse with men. Overweight (92.5%) and obese (91.5%) women are significantly *more* likely ever to have had sexual intercourse with men than normal-weight women (87.4%).

Studies of mate preference throughout the world overwhelmingly show that men prefer to mate with women with low waist-to-hip ratios *in the normal weight range.*[22] Men don't like women who are underweight, and men certainly don't like women who are overweight. So overweight and obese women could not possibly have as much sex as normal-weight women, let alone *more* sex, if men decide when and with whom to have sex. Most men would simply not choose overweight and obese women as their preferred sexual partners. Overweight and obese women can have more sex than normal-weight women only if women decide when and with whom to have sex, and men have little say in the matter.

When a man propositions a woman, she can respond in one of two ways; she can say "yes" or she can say "no." When a woman propositions a man, he can also respond in one of two ways; he can say "yes" or he can say "yes, please." He has no realistic choice to say no. Men may not be saying "yes, please" to overweight and obese women, but Kaneshiro et al.'s study clearly suggests that they are definitely saying "yes."

Because of the power of female choice, every woman has the power to predict the future, while very few men do. If a man wakes up in the morning and says to himself, "Tonight I will get laid," the prediction will fail a vast majority of times, unless he is incredibly handsome. Most young men in fact do make this prediction every morning and go to bed alone and disappointed every night. If a woman—any woman—wakes up in the morning and says to herself, "Tonight I will get laid," the prediction will likely come true every time. Such is the power of female choice.

What Does Female Choice Mean for Intelligent Men's Preference for Sexual Exclusivity?

If sex and mating were an entirely or mostly male choice, and it happened whenever and with whomever men wanted, then it would be reasonable to conclude that more intelligent men, who value sexual exclusivity more than less intelligent men, may be less likely to be sexually unfaithful than less intelligent men. However, sex and mating are a female choice.

There are several complicating factors here. First, more intelligent individuals—both men and women—are more likely to attain higher status and accumulate more resources than less intelligent individuals, at least in the evolutionarily novel environment of today.[23] And women prefer men of higher status and greater means as their mates.[24] Second, more intelligent individuals—both men and women—are on average physically more attractive than less intelligent individuals.[25]

For example, in a recent study with the NCDS data, those who are described as "attractive" by two different judges are significantly more intelligent (IQ = 104.2) than those who are described as "unattractive" (IQ = 91.8), by 12.4 IQ points! The difference was much greater among boys (105.0 vs. 91.4) than among girls (103.6 vs. 92.3).[26] In fact, by pure coincidence, intelligence in the NCDS data is just as strongly associated with physical attractiveness as it is with education; in both, the bivariate correlation coefficient is $r = .381$. Among other things, it means that, if you want to estimate someone's intelligence without giving them an IQ test, then you would do just as well to base your estimate on their physical attractiveness as you would to base it on their years of formal education![27]

At any rate, women prefer handsome men as mates, particularly for short-term mating ("casual sex" or "affairs").[28] Third, general intelligence is strongly correlated with height; more intelligent individuals—both men and women—are significantly taller than less intelligent individuals.[29] And, once again, women prefer taller men as mates.[30]

So, if you simply compare more intelligent men and less intelligent men, without statistically controlling for their social status, income, wealth, physical attractiveness, and height, I am almost certain that more intelligent men are more likely to have affairs than less intelligent men, not necessarily because they are more intelligent, but because they are more likely to have higher social status and greater resources, and to be physically more attractive and taller. If you partial out the effects of status, resources, physical attractiveness, height, and all the other potential confounds and correlates of general intelligence, then, and only then, may more intelligent men be less likely to have affairs.

However, to the best of my knowledge, no one has examined the partial effect of general intelligence on the probability of having affairs, net of all potential confounds. So we need further research in this area to determine what effect (if any) men's general intelligence has on their actual sexual behavior (extramarital or otherwise).

In fact, data from the GSS do suggest that more intelligent men (and women) are more likely to have affairs. The mean IQ of men who have had an extramarital affair is significantly (though only slightly) higher than that of men who have never

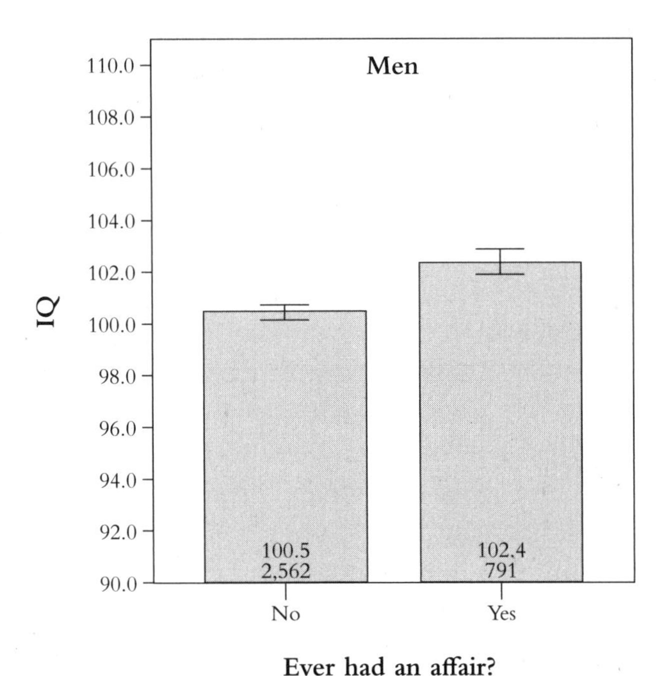

Figure 7.2 Association between intelligence and having an affair among men

had an extramarital affair (102.4 vs. 100.5). Among women, the difference is slightly larger (104.6 vs. 101.5).

The association between IQ and extramarital affairs remains significant, for both men and women, even after I control for education, income, and social class, as well as race, age, current marital status, number of children, religion, and religiosity. The effect of IQ is much stronger for women than for men. It is not clear to me why more intelligent women are more likely to have affairs than less intelligent women. Interestingly, as is quite often the case, intelligence and education have *opposite* effects on extramarital affairs for women. While more intelligent women are more likely to have affairs, more educated women are less likely to have them.

Unfortunately, the GSS does not measure the respondent's height or physical attractiveness, so I cannot control for them in

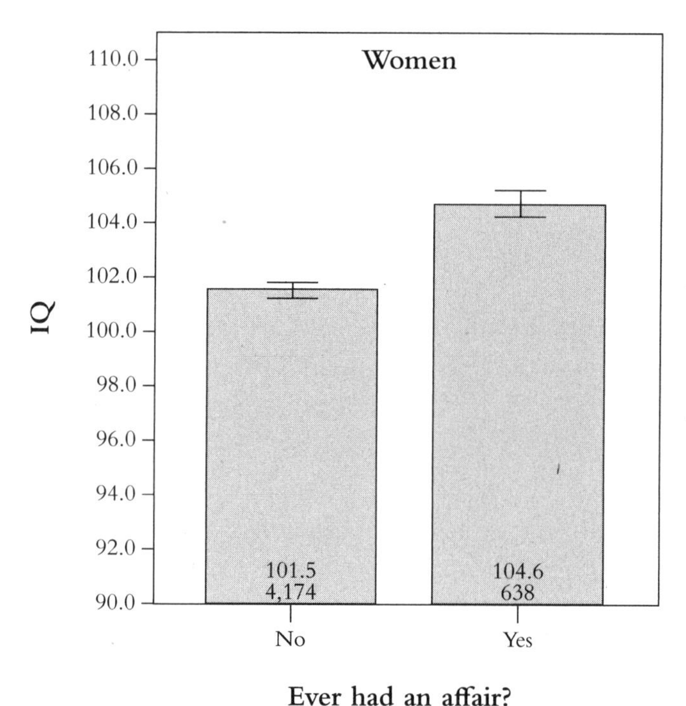

Figure 7.3 Association between intelligence and having an affair among women

the analysis. Add Health and NCDS, both of which do measure height and physical attractiveness, do not measure the respondents' experience of extramarital affairs. It therefore remains to be seen whether the significant association between intelligence and the propensity to have affairs among men is a function of the greater physical attractiveness and height of more intelligent men.

Note that the Intelligence Paradox is about individual preferences and values, what people *desire* and *want* in their heads; it's not necessarily about what people actually do. If people have complete choice over their behavior, they are expected to pursue what they desire and want, but they do not always have such complete choice. And, when it comes to sex and mating, men have very little choice.

Intelligence and Marriage Institutions

Now what does the association between intelligence and the value for sexual exclusivity among men mean for the society? What would be the macrolevel implications at the societal level of the microlevel association between intelligence and preference for sexual exclusivity among men at the individual level?

It turns out that the institution of marriage (monogamy vs. polygyny) is very strongly associated with the average intelligence of the population.[31] Across 187 nations, average intelligence and the degree of polygyny is correlated at $r = -.615$. Even net of such relevant factors and potential confounds as economic development, average level of education, geographical location, Muslim religion, and income inequality, the average intelligence of a population is very strongly associated with the level of polygyny in society. The more intelligent the population, the less polygynous (and the more monogamous) it is.

In a 1999 paper with Mary C. Still,[32] I argued that a major determinant of the level of polygyny in society was income inequality. The more unequal the income distribution, the more polygynous the society. This is because, when the income inequality among men is great, it makes more economic sense for women to share a wealthy man than to monopolize a poor man. In the memorable words of George Bernard Shaw (one of the founders of the London School of Economics and Political Science, where I teach), "The maternal instinct leads a woman to prefer a tenth share in a first rate man to the exclusive possession of a third rate one."[33] This is no longer true when the income inequality among men is less. Under more egalitarian income distribution, women should prefer "the exclusive possession of a third rate" man to "a tenth share in a first rate man."

The earlier finding still stands; income inequality does greatly increase the level of polygyny in society. However, it turns out that the average intelligence of a population is an even stronger determinant of the level of polygyny. In fact, the average intelligence of a population is the strongest determinant of it of all factors considered, even stronger than the Muslim religion. Yes, Muslim nations are more polygynous than non-Muslim nations, but the effect of the average intelligence of the population on the level of polygyny in society is much greater.

Chapter 8

Why Night Owls Are More Intelligent than Morning Larks

I first became interested in circadian rhythm—why some people are night owls while others are morning larks—when I lived in Christchurch, New Zealand, for one year. I lived on Riccarton Road, one of the main thoroughfares in Christchurch. Down Riccarton Road from where I lived, there was a supermarket called Countdown that was open 24 hours a day, 7 days a week.

I would often go to Countdown to shop at three o'clock in the morning. Every time I did, I noticed that the place was crawling with Asian customers. This was before the recent explosion of Asian immigration to New Zealand, so Asians were still a small minority in Christchurch back then. I would not see many Asians in Christchurch most of the time, except at three o'clock in the morning in a 24-hour supermarket.

I began wondering then if this was because Asians were more nocturnal than other races. I had not thought about possible race differences in circadian rhythm until I lived in New Zealand and encountered a large number of Asians at three o'clock in the morning in a 24-hour supermarket. It was many years before I solved the mystery of late-night Asian shoppers.

Choice within Genetic Predisposition

Choice is not incompatible with or antithetical to genetic influence. As long as heritability (the proportion of the variance in behavior explained by genes) is less than 1.0, genes merely set broad limits, and individuals can still exercise some choice within broad genetic constraints.

For example, political scientists in the emerging field of genopolitics[1] have discovered that two genes are responsible for predisposing individuals to be more or less likely to vote in elections.[2] In other words, whether you choose to turn out to vote in any given election is partially determined by your genes. However, individuals can still choose to turn out to vote or not for any election, and there are environmental (nongenetic) factors that can predict their voting, such as whether the candidate you voted for in the last election won or lost.[3] You become more likely to vote if you voted for a candidate who won in the last election, and less likely to vote if you voted for a candidate who lost in the last election.

Similarly, genetic influences and constraints do not preclude individual acquisition of values and preferences. Individuals can still choose certain values and preferences even in the face of genetic predisposition. In fact, Turkheimer's first law of behavior genetics,[4] which I briefly introduce in Chapter 3, states that all human traits are heritable and influenced, at least in part, by genes. So if genetic predisposition is incompatible with personal choice, it would mean that humans have absolutely no choice to make at all. In reality, however, humans make choices *in the face of* genetic predisposition.

For example, both political ideology[5] and religiosity[6] have now been shown to have genetic bases. Some individuals are genetically predisposed to be liberal or conservative, or more or less religious. Yet, as we see in Chapters 5 and 6, more intelligent children are more likely to grow up to be liberal and less likely to grow up to be religious.[7] Similarly, whether you are a night person or a morning person is partly genetically determined, but that does not mean that people still cannot consciously and volitionally choose to be a night owl or a morning lark.

Night Life Is Evolutionarily Novel

Virtually all species in nature, from single-cell organisms to mammals, including humans, exhibit a daily cycle of activity called the circadian rhythm. "This timekeeping system, or biological 'clock,' allows the organism to anticipate and prepare for the changes in the physical environment that are associated with day and night, thereby ensuring that the organism will 'do the right thing' at the right time of the day."[8] The circadian rhythm in mammals is regulated by two clusters of nerve cells called the suprachiasmatic nuclei (SCN) in the anterior hypothalamus.[9]

Geneticists have by now identified a set of genes that regulate the SCN and thus the circadian rhythm among mammals.[10] A behavior genetic study of 977 South Korean twin pairs shows that the heritability in morningness-eveningness (whether you are a morning person or a night person) is .45 and nonshared environment accounts for 55% of the variance, while shared environment does not appear to explain any of the variance.[11] So the circadian rhythm appears to be yet another human trait that roughly follows the 50–0–50 rule I discuss in Chapter 3.

"For most animals, the timing of sleep and wakefulness under natural conditions is in synchrony with the circadian control of the sleep cycle and all other circadian-controlled rhythms. *Humans, however, have the unique ability to cognitively override their internal biological clock and its rhythmic outputs.*"[12] While there are some individual differences in their circadian rhythm, where some individuals are more nocturnal than others, humans are basically a diurnal (day-living) species, as are all extant monkey and ape species, except for one.[13]

Humans rely very heavily on vision for navigation but, unlike genuinely nocturnal species, cannot see in the dark or under little lighting. Our ancestors did not have artificial lighting during the night until the domestication of fire. Any human in the ancestral environment up and about during the night would have been at risk of predation by nocturnal predators. It is therefore safe to assume that our ancestors rose at around dawn and went to sleep at around dusk, to take full advantage of the natural light provided by the sun, and that "night life" (sustained and organized activities at night after dark) is probably evolutionarily novel.

In order to ascertain the extent to which our ancestors might have engaged in nocturnal activities, I have once again consulted ethnographic records of traditional societies throughout the world. In the 10-volume compendium *The Encyclopedia of World Cultures*,[14] which extensively describes *all* human cultures known to anthropology, there is no mention of nocturnal activities in any of the traditional cultures. There are no entries in the index for "nocturnal," "night," "evening," "dark(ness)," or "all-night." The few references to the "moon" are all religious in nature, as in "moon deity," "Mother Moon (deity)," and "moon worship." The only exception is the "night courting," which is a socially approved custom of premarital sex observed among the Danes and the Finns, which are entirely western cultures far outside of the ancestral environment.

In addition, I have consulted the same five extensive (monograph-length) ethnographies that I consult in Chapters 5 and 6.[15] Many of these ethnographies contain a section where the authors describe what usually happens and what people routinely do in a typical day in the tribal society under study.

These detailed ethnographic records make it clear that the day for people in these traditional societies begins shortly before sunrise, and ends shortly after sunset. "Daily activities begin early in a Yanomamö village."[16] "The day begins about 6 a.m., when the sun is about to rise."[17] The only routine activities conducted after dark are people conversing and visiting with each other as they drift off to sleep. "Despite the inevitable last-minute visiting, things are usually quiet in the village by the time it is dark."[18] "Most evenings are spent quietly chatting with family members indoors. If the moon is full then it is possible to see almost as well as during the day, and people take advantage of the light by staying up late and socializing a great deal."[19] "After cooking and consuming food, evening is often the time of singing and joking. Eventually band members drift off to sleep, with one or two nuclear families around each fire."[20] The only nocturnal activities, other than chatting visiting, and making speeches, that I can find in all of these ethnographies is when Mukogodo men go searching for missing animals in the dark, *if* one happens to be missing.[21]

It may be significant in this context that humans mostly evolved in sub-Saharan Africa near the Equator, where the length of the day remains more or less constant throughout the year. So the practice of "wake up at dawn, go to sleep at dusk" would have produced days of roughly the same length throughout the year. At higher latitude, however, the same practice would have produced days of varying lengths, longer in the summer and shorter in the winter. In the extreme cases, near the arctic circles, days would

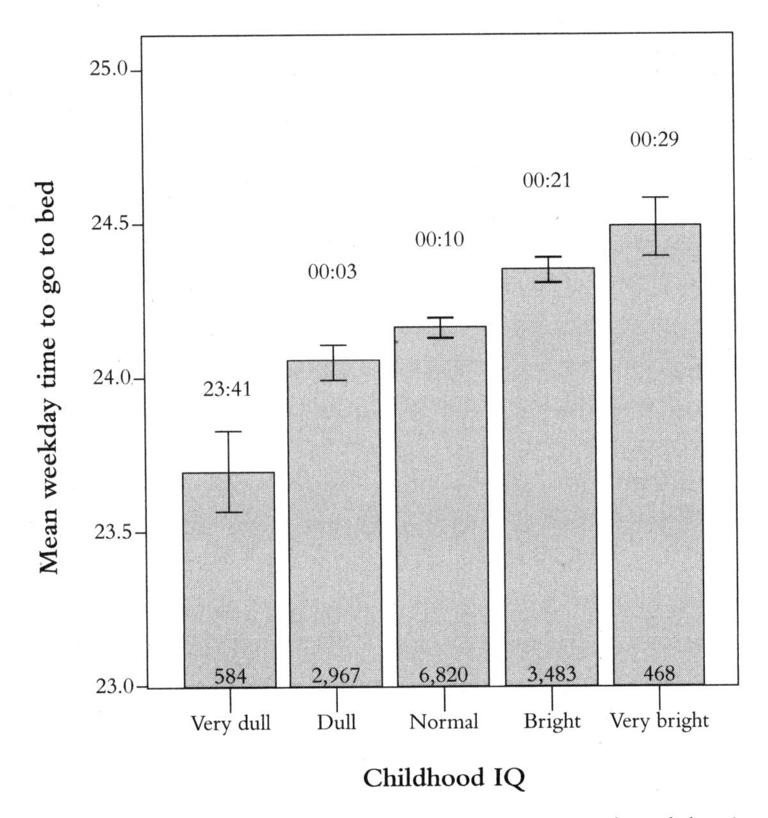

Figure 8.1 Association between childhood intelligence and weekday time to go to bed

never have ended in the summer and would never have begun in the winter. Thus humans in higher latitudes would have had to wake up before dawn and stay up after dusk if they wanted to have days of roughly the same length throughout the year. Incidentally, the average intelligence of populations tends to be higher at higher latitudes (and longitudes), even controlling for the average temperature.[22]

Ethnographic evidence of traditional societies therefore suggests that our ancestors probably had a largely diurnal lifestyle, and sustained and routine nocturnal activities may be evolutionarily novel. The Intelligence Paradox would therefore predict that

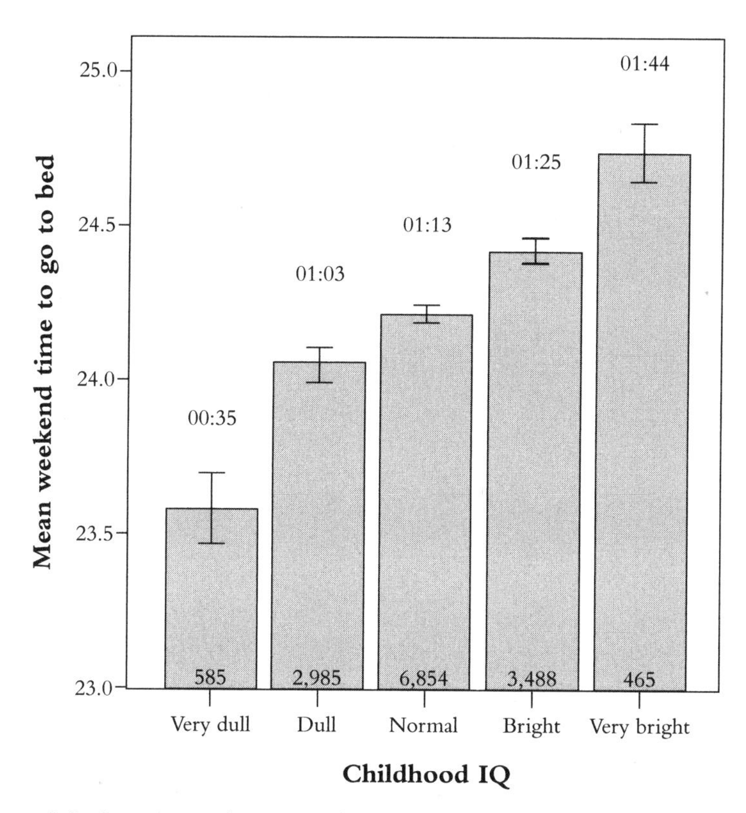

Figure 8.2 Association between childhood intelligence and weekend time to go to bed

more intelligent individuals are more likely to be nocturnal, getting up later in the morning and going to bed later in the evening, than less intelligent individuals.

Previous to my 2009 article with Kaja Perina,[23] there had only been one study which examined the association between intelligence and circadian rhythm.[24] The 1999 study found that, in a small sample of US Air Force recruits, evening types were significantly more intelligent than morning types. This is consistent with the prediction of the Intelligence Paradox.

The analysis of the Add Health data confirm this prediction of the Intelligence Paradox. Net of age, sex, race, marital status,

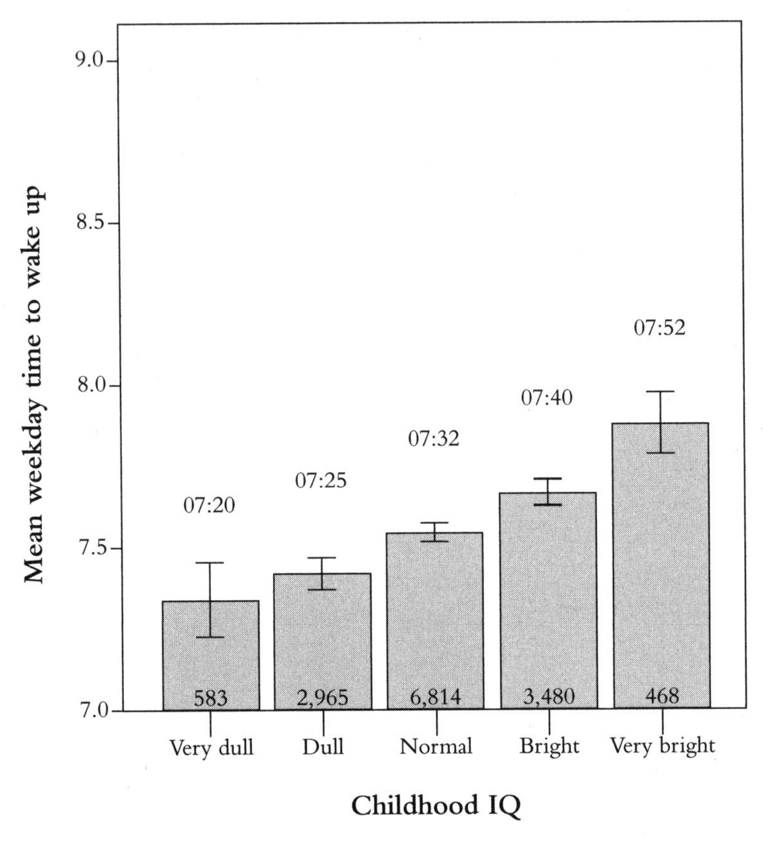

Figure 8.3 Association between childhood intelligence and weekday time to wake up

parental status, education, income, religion, whether currently in school, and the number of hours worked, more intelligent children are more likely to grow up to be nocturnal in their early adulthood. More intelligent individuals go to bed later, both on weeknights and on weekend nights, and they wake up later on weekdays (but not on weekends, for which the positive effect of general intelligence on nocturnality is not statistically significant).

Figures 8.1 through 8.4 show that the association between childhood IQ and adult nocturnality is monotonically positive, even though absolute differences are not very large. For example,

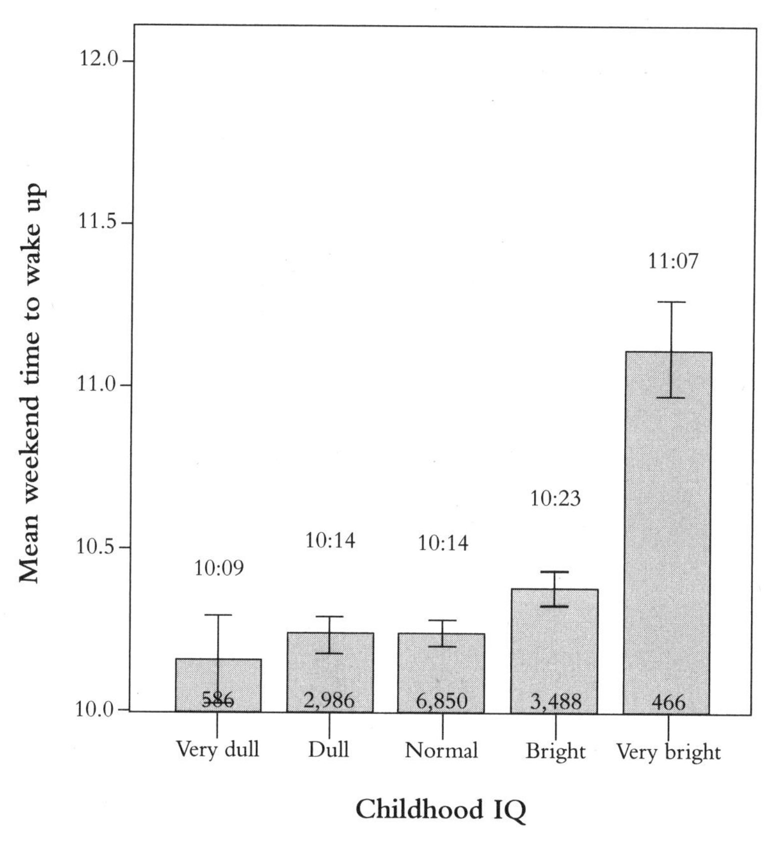

Figure 8.4 Association between childhood intelligence and weekend time to wake up

on weeknights, "very dull" children (with IQs below 75) on average go to bed at 23:41 in early adulthood, whereas "very bright" children (with IQs above 125) on average go to bed at 00:29. In general, the more intelligent they are in junior high and high school, the later they go to bed and the later they wake up in early adulthood. The probability that one would get the patterns as strong as those represented in the four figures above purely by chance, when there is actually no association between childhood intelligence and adult circadian rhythm, is one in 10,000 or smaller.

So Are Asians Really More Nocturnal than Others?

So are Asians really more nocturnal than other races, as was my impression on my frequent trips to Countdown in Christchurch at three o'clock in the morning? I don't have data on Kiwis, but Add Health data on Americans do suggest that they are.

Compared to all other races (whites, blacks, and Native Americans), Asians go to bed significantly later on weeknights (00:43 vs. 00:08) and on weekend nights (01:27 vs. 01:12). However,

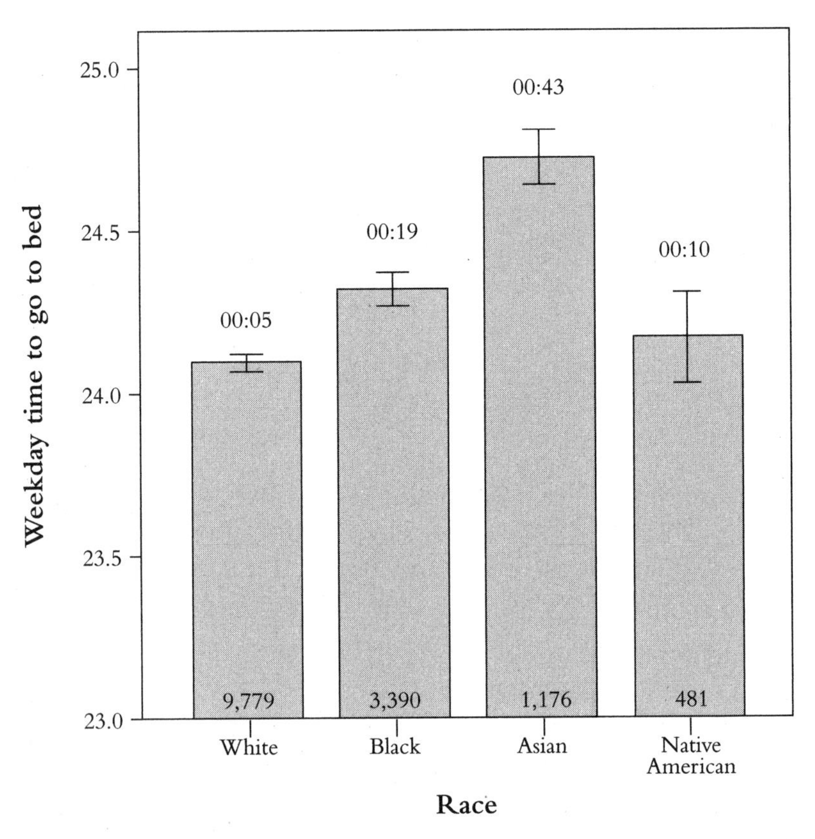

Figure 8.5 Association between race and weekday time to go to bed

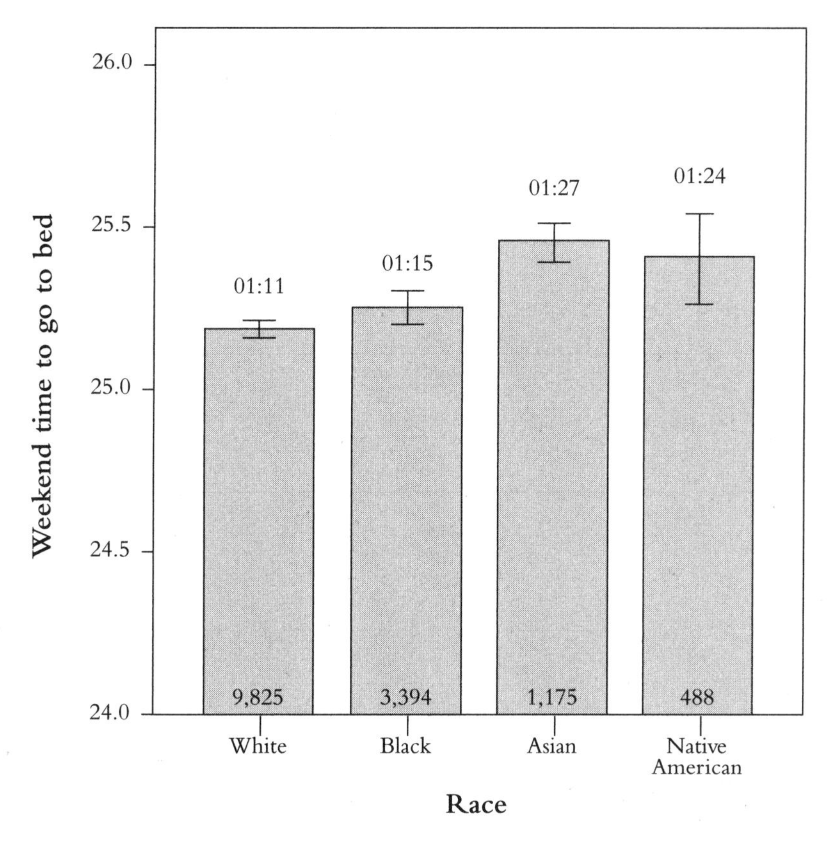

Figure 8.6 Association between race and weekend time to go to bed

Asians do not wake up significantly later than others either on weekdays (07:38 vs. 07:32) or on weekends (10:20 vs. 10:18). In general, races differ significantly from each other in what time they go to bed, both on weeknights and weekend nights, but they do not differ significantly from each other in what time they wake up, either on weekdays or on weekends.

So it appears that my casual observations at Countdown in Christchurch so many years ago may have some empirical basis. Compared to other races, Asians do appear to stay up (and, presumably, shop for groceries, among other things) later into the night. The bivariate associations between being Asian and

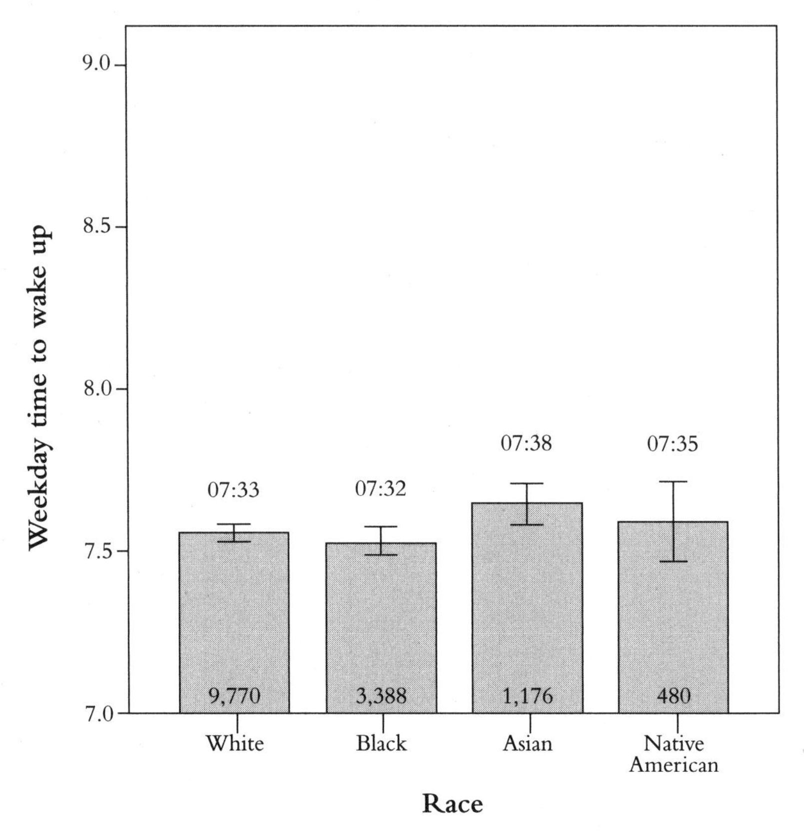

Figure 8.7 Association between race and weekday time to wake up

nocturnality (what time they go to bed at night), represented in Figures 8.5 and 8.6, remain statistically significant even when childhood IQ is controlled for. Even net of childhood intelligence, Asians go to bed significantly later than others every night. So it is *not* because they are slightly more intelligent than other races that Asians are more nocturnal.

However, the association disappears once I control for all the other social and demographic variables included in the multiple regression analyses (age, sex, current marital status, parental status, education, income, religion, whether currently in school, and the number of hours worked). In fact, the only significant effect of

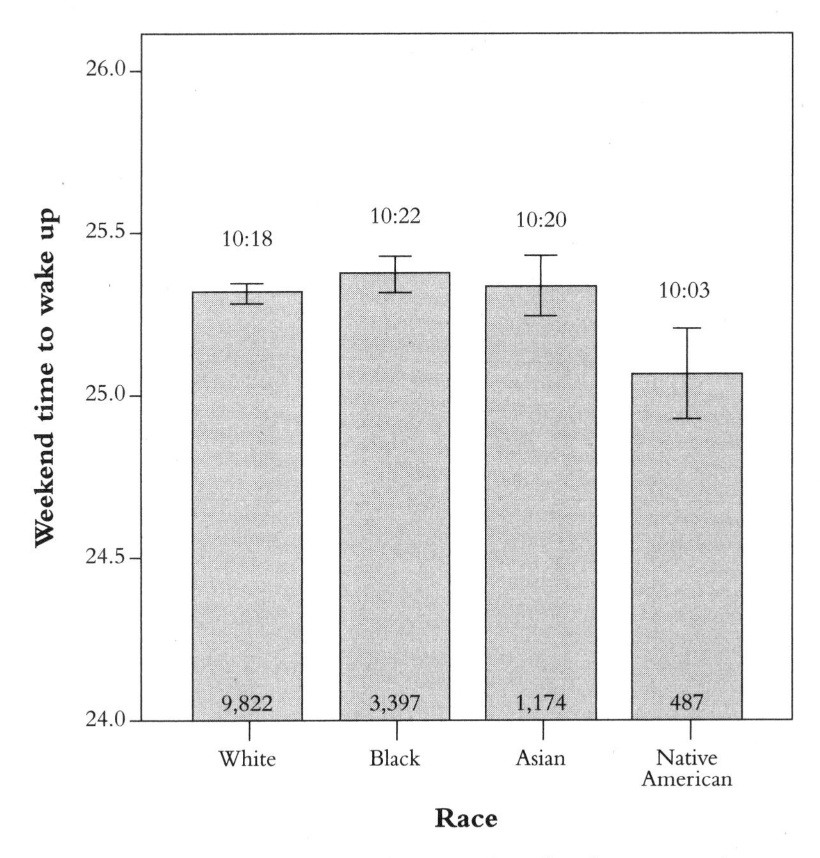

Figure 8.8 Association between race and weekend time to wake up

race in the multiple regression analyses is being black. Compared to whites (the reference category), blacks go to bed later on weekends (but not on weeknights) and wake up later on weekends (but not on weekdays).[25]

Unfortunately, the Intelligence Paradox is completely mute on genuine race differences in nocturnality (or anything else) net of general intelligence. So I do not know why Asians or blacks stay up later into the night or wake up later in the morning.

Chapter 9

Why Homosexuals Are More Intelligent than Heterosexuals

Recall the discussion of the naturalistic and moralistic fallacies in Chapter 1. Science is all about what is, and it is never about what ought to be. Nothing in science is good or bad; everything just is (or isn't). Science does not make any moral judgments.

Even though some form of homosexuality is observed in many species,[1] the basic biological design of all mammalian species is heterosexual reproduction, and exclusive or predominant homosexuality is rare in nature. Most importantly, we are not descended from ancestors who were exclusively homosexual, so it is unlikely that homosexuality has been part of human nature throughout evolutionary history. In this narrow sense, being homosexual is unnatural; humans (as all the other species) are not designed to be homosexual.

So the Intelligence Paradox would predict that more intelligent individuals may be more likely to be homosexual than less intelligent individuals. But before we discuss the potential effect of general intelligence on sexual behavior, let's first of all talk about what it means to be homosexual or heterosexual. It's not as straightforward as you may think at first.

What Does It Mean to Be Homosexual?

Heritability of male homosexuality—the proportion of the variance in male sexual orientation that is explained by genetic factors—has been estimated to be between .26[2] and .60.[3] So somewhere between a quarter and two-thirds of the variance in male sexual orientation is genetically determined. The remaining variance, however, appears to be accounted for by prenatal exposure to androgen in the womb; the more androgen—male hormones—the male fetus is exposed to in the womb during gestation, the more likely they are to become homosexual.[4] This is why, for example, men who have more older brothers are more likely to be homosexual.[5] Each additional brother increases the odds that a man becomes homosexual by 33–38%. The current consensus among sex researchers is that, between genes and prenatal hormones, men's sexual orientation is largely determined before birth,[6] while women's sexual orientation is more malleable and fluid.[7]

There are four different measures of sexual orientation:[8]

1. Self-identified labels = whether you consider yourself to be homosexual, bisexual, heterosexual, etc.

2. Actual sexual behavior = with whom you have sex.
3. Self-reported sexual feelings = to whom you are sexually attracted and about whom you have sexual fantasies.
4. Genital or brain responses = physiologically measured arousal to male or female images.

In their 2005 book *Born Gay*, the sex researchers Glenn Wilson and Qazi Rahman note that self-identified labels can be influenced by political and cultural climate. For example, many homosexuals throughout history and in some oppressive regimes today have been forced to remain in the closest due to social pressure and threat of legal punishment, including death. Actual sexual behavior, Wilson and Rahman note, can be influenced by opportunities and circumstances. For example, many heterosexual men often have sex with other men while in prison due to the complete absence of potential female sexual partners.

In contrast, sexual feelings and physiological measures (3 and 4 above) are more stable and closer to individuals' "true" sexual orientation. For example, self-identified heterosexual men who are openly and publicly homophobic may nonetheless show genital response of arousal to sexual images of other men.[9] (Yes, some of the most openly homophobic men turn out to be secretly homosexual themselves. Remember the ex-Marine neighbor of Kevin Spacey's character in *American Beauty*?) Wilson and Rahman also note that homosexual fantasies are quite common in heterosexual men and women as a form of "mental explorations"[10] and that measuring homosexuality with reported sexual fantasies and desires assumes that survey respondents are completely honest about them.

All in all, Wilson and Rahman conclude that physiologically measured arousal (genital or brain responses to sexual images of men or women) is probably the most accurate measure of *true* sexual orientation, and the other three measures may correlate poorly with it and may deviate from their true sexual orientation, especially among women.[11]

Given that an individual's true sexual orientation, at least for men, is prenatally determined (either genetically at the moment of conception or through prenatal exposure to androgen during gestation prior to birth), it is not likely that more intelligent individuals are more likely to be *truly* homosexual. There is a possibility, however, that the (as yet undiscovered) genes for intelligence are somehow linked to the (as yet undiscovered) genes for homosexuality in men, because genes for both intelligence and male homosexuality appear to be located on the chromosome Xq28.[12] At any rate, given that the first three measures of sexual orientation (1, 2, and 3 above) are more malleable and subject to conscious choice and self-presentation, it may also be possible that more intelligent individuals are more likely to appear homosexual by these measures, that is, *if* homosexual identity and behavior are evolutionarily novel. Regardless of their *true* sexual orientation, more intelligent individuals may identify themselves as homosexual, engage in homosexual behavior, or report homosexual fantasies and desires.

Evolutionary Novelty of Homosexual Identity and Behavior

In order to examine the extent to which our ancestors might have identified themselves as homosexuals and engaged in homosexual behavior, I have once again consulted the same ethnographic records of traditional societies throughout the world that I use in earlier chapters. When it comes to homosexuality, contemporary hunter-gatherers, while not exactly the same as our ancestors, are probably a lot more similar to our ancestors than are residents of San Francisco or Brighton today.

The 10-volume compendium *The Encyclopedia of World Cultures*[13] mentions male homosexuality in seven different cultures (Foi, Gebusi, Kaluli, Keraki, Kiwai, Marind-anim, and Sambia). However, these are phylogenetically closely related tribes all in Papua New Guinea, and all practices of male homosexuality in these Papua New Guinean cultures occur strictly as part of initiation rites for boys.

For example, "Gebusi believe boys must be orally inseminated to obtain male life force and attain adulthood. Insemination continues during adolescence and culminates in the male initiation (*wa kawala*, or 'child becoming big') between ages 17 and 23."[14] And among the Sambia, "male maturation requires homoerotic insemination to attain biological competence. Initiation rituals thus involve complex homosexual contact from late childhood until marriage, when it stops."[15]

These homosexual practices in Papua New Guinea appear highly ritualized and culturally mandated. There appears very little individual choice involved, and, as such, homosexuality does not appear to be an individual-difference variable (where some people practice it while others don't, where some people are homosexual and others are heterosexual). It therefore appears quite different from what we normally mean by "sexual relations," which involve choice, emotions, and attachment. At any rate, it is very difficult to suggest that homosexuality was a routine part of our ancestors' life if its present-day practice on a large scale among traditional societies is limited only to one island in the South Pacific far outside of the ancestral environment of sub-Saharan Africa.

In addition, I have also consulted the five extensive (monograph-length) ethnographies of traditional societies around the world which I rely upon in earlier chapters.[16] In any of these ethnographies, there is no mention of explicit homosexual relationships among the members of the societies under study. The only potential exception is the panegi among the Ache.[17]

> Some men in our sample never had any children and others never acquired a wife. One category of men in Ache society opts out of the male mating pool altogether. These men, called *panegi*, take on a female socioeconomic role (the world *pane* means unsuccessful or unlucky at hunting). Men who are *panegi* generally do not hunt, but instead collect plant resources and insect larvae. They weave baskets, mats, and fans, and make tooth necklaces, bowstrings, and other female handicrafts. They spend long hours cooking, collecting firewood or water, and caring for children. *Most informants stated that "panegis" did not ever engage in homosexual behavior (oral or anal) prior to first contact.* A few informants said they were not sure, but had never heard of such behavior.

Panegis are apparently small in stature.[18] And, at least in North America, homosexual men are shorter than heterosexual men.[19] So perhaps the panegis among the Ache might have been genetically and hormonally predisposed to homosexuality. But the ethnographic records make it clear that they nonetheless did not engage in homosexual behavior prior to first contact with the western civilization.

It is very important to point out, however, that even very extensive ethnographies, based on long-term fieldwork by very experienced anthropologists familiar with the local culture and language, may not always detect instances of homosexuality. This may especially be the case if homosexuality is condemned and negatively sanctioned in the local culture. So the absence of references to homosexuality in these ethnographies is not by itself conclusive evidence of its absence in traditional societies.

However, the same ethnographers and anthropologists have nonetheless been adept at uncovering evidence of other negatively sanctioned and concealed behavior like murder, theft, infanticide, and extramarital affairs in the same traditional societies. So the near total absence of any documentation of homosexual behavior as an individual choice may suggest that it may be relatively rare in such societies. It may also suggest that widespread practice of homosexual behavior may have been rare in the ancestral environment, and it may therefore be evolutionarily novel.

If homosexual identity and behavior are evolutionarily novel, then the Intelligence Paradox would predict that, regardless of their *true* sexual orientation, more intelligent individuals may be more likely to identify themselves as homosexual, report homosexual feelings and desires, and engage in homosexual behavior than less intelligent individuals. All three main data sets for this book (the GSS, Add Health, and the NCDS) allow me to examine the association between intelligence and homosexuality.

Intelligence and Homosexuality

Add Health

Add Health uses two different measures of homosexuality. The first question asks the respondents to identify their sexual orientation as either 1 = "100% heterosexual (straight)," 2 = "mostly heterosexual (straight), but somewhat attracted to people of your own sex," 3 = "bisexual—that is, attracted to men and women equally," 4 = "mostly homosexual (gay), but somewhat attracted to people of the opposite sex," and 5 = "100% homosexual (gay)." This measure of sexual orientation corresponds to the "self-identified label" definition of it (1 above).

The analysis of Add Health data show that, even net of sex, age, race, marital status, parenthood, education, income, and religion, more intelligent children are more likely to identify themselves as homosexual in early adulthood than less intelligent children.[20] The more intelligent Add Health respondents are in junior high and high school, the more homosexual they identify themselves to be in their 20s. The effect of childhood intelligence on adult homosexual identify does not differ for men and women.

Even though childhood intelligence and education are naturally positively associated (the more intelligent they are in childhood, the greater education they attain by early adulthood), intelligence and education have opposite effects on adult homosexual identity. While more intelligent individuals are more likely to identify themselves to be homosexual, the more educated individuals are less likely to do so.

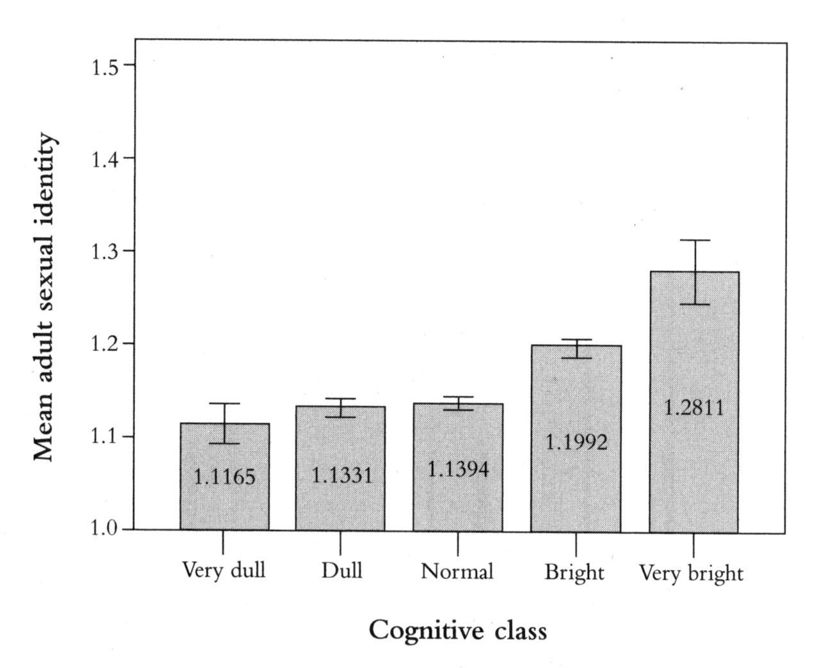

Figure 9.1 Association between childhood intelligence and adult sexual identity

Figure 9.1 represents the bivariate association between childhood intelligence and adult sexual identity. It shows that the association is monotonically positive; "very bright" children grow up to become more homosexual in their identity than "bright" children, who in turn grow up to become more homosexual than "normal" children, etc. The probability that one would observe a pattern as strong as the one depicted in Figure 9.1 purely by chance, when there is actually no association between childhood intelligence and adult sexual identity, is less than one in a hundred billion!

The second question asks, "Have you ever had a romantic attraction to a member of the same sex?" The respondents can answer either yes or no. This measure corresponds to the "self-reported sexual feelings" definition of sexual orientation (3 above).

The analysis of Add Health data shows that, net of the same factors as above, more intelligent children are more likely to have experienced adult homosexual attraction than less intelligent children.[21] The more intelligent Add Health respondents are in junior high and high school, the more likely they are to have ever experienced romantic attraction to members of the same sex. If you increase childhood intelligence by 15 IQ points (one standard deviation), then you increase the odds of expressed adult homosexual attraction by 27%.

The effect of childhood intelligence on adult homosexual attraction is significantly stronger for women than for men. In fact, underscoring their more fluid sexuality,[22] women have more than 50% greater odds of having ever experienced romantic attraction to members of the same sex than men do.

Figure 9.2 represents the bivariate association between childhood intelligence and expressed adult homosexual attraction. It shows that the association is monotonically positive. "Very bright" children are more likely ever to have experienced adult

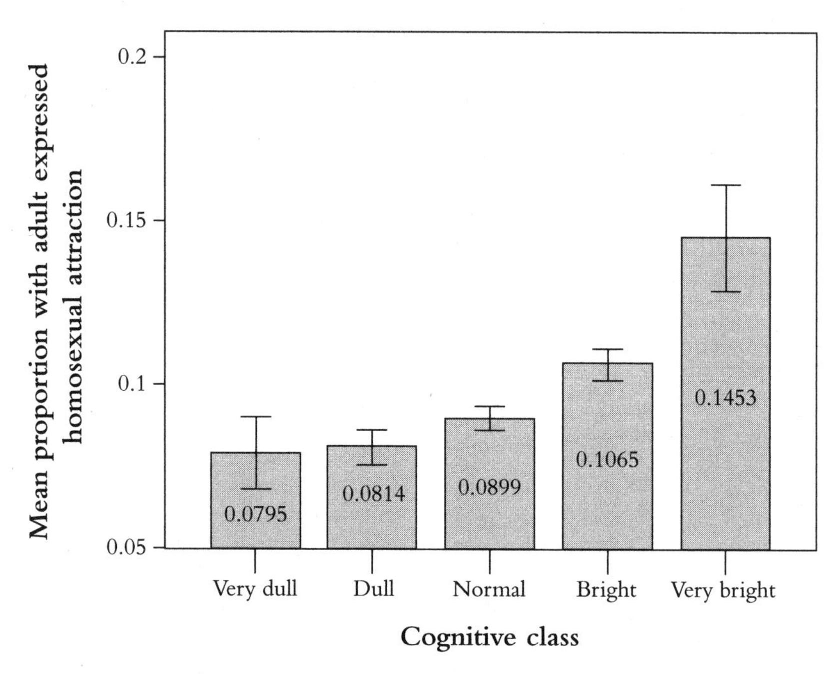

Figure 9.2 Association between childhood intelligence and expressed adult homosexual attraction

homosexual attraction than "bright" children, who are in turn more likely ever to have experienced it than "normal" children, etc. In fact, "very bright" children are nearly twice as likely to grow up to experience adult homosexual attraction as "very dull" children. The probability that one would observe a pattern as strong as the one depicted in Figure 9.2 purely by chance, when there is actually no association between childhood intelligence and expressed adult homosexual attraction, is less than one in 10,000.

GSS

While Add Health has precise measures of homosexual identity and feelings (corresponding to 1 and 3 in the list of definitions of sexual orientation above), it unfortunately lacks any measure of actual sexual behavior with members of the same sex; it only measures heterosexual sexual behavior. I therefore now turn to the GSS, which measures both homosexual and heterosexual behavior.

The GSS measures the respondents' homosexual and heterosexual behavior by asking how many sex partners of each sex they have ever had since they were 18. This measure of sexual orientation corresponds to the "actual sexual behavior" definition of it (2 in the above list).

The analysis of the GSS data shows that, consistent with the Intelligence Paradox, net of sex, age, race, social class, education, income, marital status, number of children, religion, and survey year, more intelligent individuals have more homosexual partners in their adult life than less intelligent individuals.[23] Contrary to the prediction of the Intelligence Paradox, the GSS data also show that, net of the same control variables, more intelligent individuals have more heterosexual partners in their adult life as well than less intelligent individuals.

However, the effect of intelligence on the number of homosexual partners is twice as strong as its effect on the number of heterosexual partners. As Figures 9.3 and 9.4 below show, "very bright" Americans have had eight times as many homosexual partners as "very dull" Americans (2.42 vs. .31). In sharp contrast, "very bright" Americans have had less than 40% more heterosexual partners than "very dull" Americans (9.79 vs. 7.10). In fact, "bright" Americans have had more heterosexual partners (9.98) than "very bright" Americans.

NCDS

Add Health and GSS very precisely measure homosexuality by the three more malleable, less stable definitions of sexual orientation (1, 2, and 3 in the list above). And the data show that all three

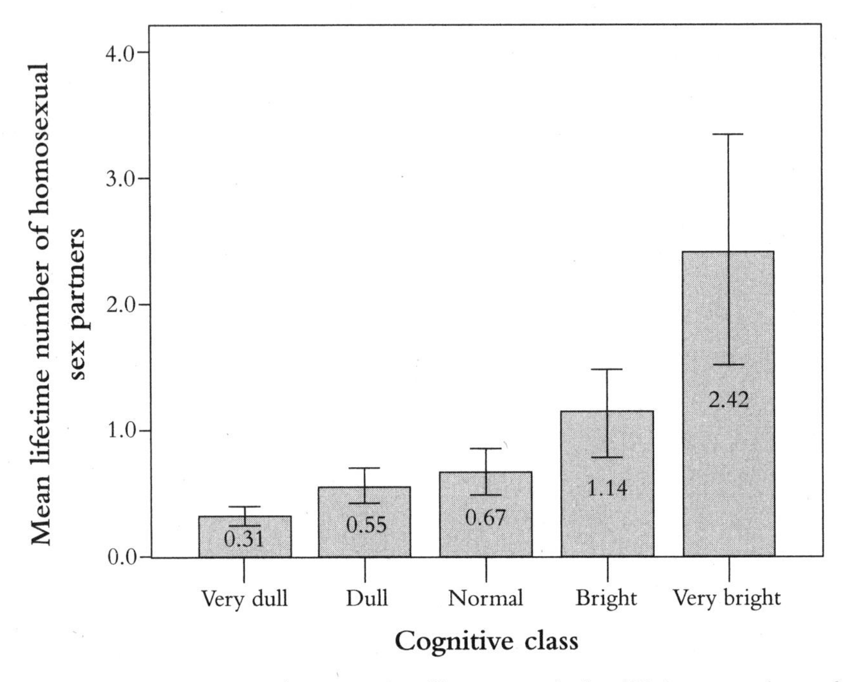

Figure 9.3 Association between intelligence and the lifetime number of homosexual partners

measures of homosexuality are significantly positively associated with intelligence. The more intelligent the individuals, the more homosexual they are, even net of a large number of potential confounds and correlates of intelligence.

However, Add Health and GSS have one small problem, as I mention in the Introduction when I discuss the details of the data sets. Both Add Health and GSS have measures of verbal intelligence, not general intelligence. While verbal intelligence is very strongly and significantly correlated with general intelligence—in fact, it is an important component of general intelligence—it is not exactly the same as general intelligence. NCDS rectifies this problem, as it has a very good and highly reliable measure of general intelligence, assessed by 11 cognitive tests administered at three different ages.

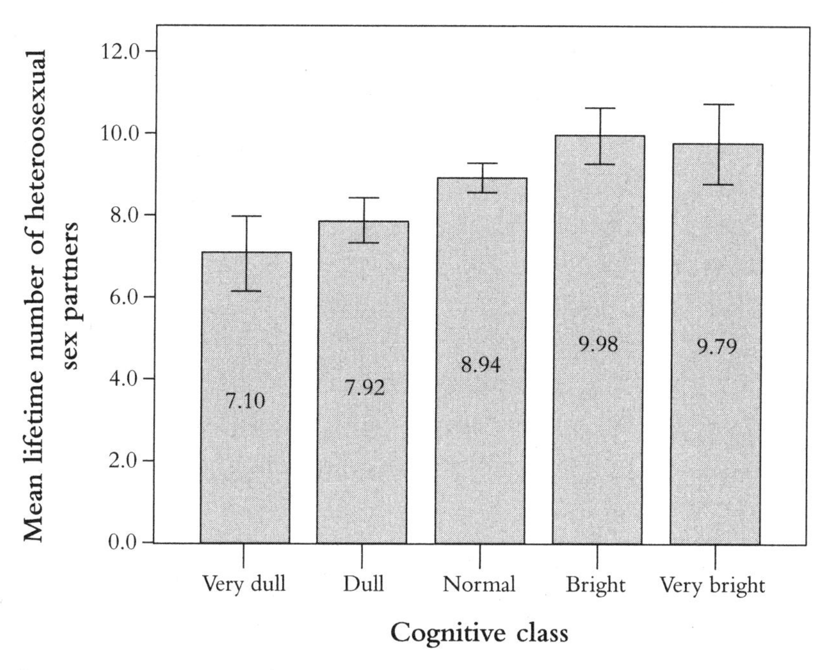

Figure 9.4 Association between intelligence and the lifetime number of heterosexual partners

Unfortunately, the only measure of sexual orientation that NCDS has is the number and sex of cohabitation partners. At age 47, NCDS asks its respondents how many same-sex and opposite-sex cohabitation partners they have had, defined as someone with whom the respondents have lived "as married" and shared an accommodation for six months or longer.

Using this measure of sexual orientation, the analysis of the NCDS data show that more intelligent children (before the age of 16) have significantly more lifetime homosexual cohabitation partners 30 years later than less intelligent children, even after statistically controlling for sex, whether currently married, whether ever married, whether ever a parent, education, income, and religion.[24] In sharp contrast, childhood general intelligence is not at all associated with the lifetime number of heterosexual partners. Since heterosexual cohabitation is eminently evolutionarily familiar, this is once again perfectly consistent with the prediction of the Intelligence Paradox.

The analyses of all three data sets (Add Health, GSS, and NCDS) uniformly confirm the prediction of the Intelligence Paradox. More intelligent children are more likely to grow up to identify themselves as homosexual and to have ever experienced romantic attraction to members of the same sex. More intelligent individuals have had more lifetime homosexual sex partners than less intelligent individuals (although intelligence is also associated with the lifetime number of heterosexual partners). More intelligent children grow up to have a larger number of lifetime homosexual cohabitation partners 30 years later, but childhood general intelligence does not predict the lifetime number of heterosexual partners. The positive association between intelligence and homosexuality appears to be quite strong and robust.

Chapter 10

Why More Intelligent People Like Classical Music

I first became interested in the possible effect of general intelligence on musical tastes when I was visiting my wife's hometown of Novgorod, Russia, in June 2002. Novgorod is an ancient provincial town, not at all cosmopolitan like Moscow or St. Petersburg. And it was years before the current influx of guest workers from the former Soviet Republics in Central Asia into Novgorod and other Russian cities. So I was just about the only Asian—the only non-Slav—in the entire town of Novgorod, and I stuck out like a sore thumb everywhere I went. People stared at me because it was obvious to all that I was not a local.

The only other time I experienced anything like that in my life was when my car—my trusted 1977 Datsun Cherry F-10 hatchback—broke down in Wallace, Idaho, during my transcontinental drive in the early fall of 1986. I had to spend several hours in Wallace, while my car was being repaired in a local garage. I appeared to be the only non-white person in the entire town of Wallace that day (or, quite possibly, any day), and everybody looked at me like I was a rock star. A group of teenagers would walk by and wave at me simply because I looked different.

Anyway, being in Novgorod was like being in Wallace, Idaho, all over again, and I was the only Asian in the whole town. I did not see another non-Slavic face during my entire visit.

One night, my wife and I decided to go to a classical music concert held at a local concert hall. It was a very small concert, with a small audience and a small orchestra. Yet there she was, the first violin of the small local orchestra in Novgorod was an Asian woman. She was the only Asian I saw in all of Novgorod during my entire visit.

Could this be a coincidence? I don't know anything about classical music, but casual observations seemed to suggest that many of the famous classical musicians throughout the world were either Jewish or Asian, the two ethnic groups with the highest average intelligence. It also seemed to me that many of the people who enjoyed listening to classical music (which decidedly does not include me) were typically highly educated and upper-class (therefore, more intelligent) people. Could there possibly be a connection between intelligence and appreciation for classical music? Are more intelligent people more likely to appreciate (and therefore perform) classical music? If so, why? What's special about classical music?

Here's another casual observation. If you like driving across the country in the United States (as I do), and if you are a fan of the National Public Radio (as I am), you may notice a pattern in the character of NPR stations across the country. In big cities (like New York or Washington DC), NPR stations tend to be news and talk stations and air news and talk programs throughout the day. In small towns, in contrast, NPR stations tend to air news and talk programs (like *Morning Edition*, *All Things Considered*, and

Fresh Air) only during the morning and evening driving hours, but otherwise air music throughout the day and night. I assume this is because carefully produced news and talk programs are more expensive to purchase and air than CDs, and NPR stations in large cities have larger budgets that allow them to purchase such programs, whereas NPR stations in smaller towns with smaller operating budgets must play music from CDs most of the day.

Typically, NPR and other radio stations which play music have "themes." No radio stations play a random collection of music; they usually focus on certain genres of music to play. So there are "classic rock" stations, and there are "country western" stations. Over the years, I have noticed that NPR stations that are not news and talk stations frequently play classical and jazz music.

According to surveys conducted on behalf of the Recording Industry Association of America in 2008,[1] only 1.9% of a representative sample of Americans purchased classical music (in all formats) in the past month, and 1.1% purchased jazz. The largest number of people (31.8%) purchased rock music in the past month. More than 10 times as many people purchased rock music than classical and jazz combined. In sharp contrast, of the 143 NPR stations which provide online streaming as of April 2011 (out of a total of 910 NPR stations nationwide[2]), 42.7% (61) play classical music, and a further 14.0% play jazz. Only 30.8% play the combined genres of rock, pop and folk.[3] Judging by these statistics, classical and jazz listeners are nearly 20 times overrepresented among the NPR station listeners (3% of the American population vs. 56.7% of NPR stations).

But why is this? Why do NPR listeners like to listen to classical or jazz music? NPR stations and their listeners are notoriously and overwhelmingly left-wing liberals. And, as I show in Chapter 5, left-wing liberals are on average more intelligent than right-wing conservatives. Does that mean that more intelligent radio listeners are more likely to prefer classical or jazz music? If so, why?

In order to answer these questions, I first had to find out how music initially came about in human evolutionary history. What is the evolutionary origin of music? Why are humans musical?

Evolutionary Origins of Music

In comparison to evolutionary origins and functions of language and art, anthropologists and archeologists have paid scant attention to the origin of music. In his book *The Singing Neanderthals: The Origins of Music, Language, Mind and Body*,[4] the cognitive archeologist Steven Mithen offers a novel theory of the evolution of music. Mithen argues that language and music had a common precursor—called musilanguage[5]—which later developed into two separate systems of music and language.

There are two distinct perspectives on the evolution of language. The *compositional approach*[6] suggests that words came before sentences. A lexicon of words that referred to specific entities like "meat," "fire," and "hunt" emerged first, and were later combined into phrases, and then into sentences. Grammar emerged at the end to dictate how words could be combined into sentences.

In contrast, the *holistic approach*[7] proposes that sentences came before words. It suggests that the precursor to human language was a communication system composed of messages in the form of arbitrary strings of sounds rather than words. Each individual utterance or sequence of sounds was associated with a specific meaning. These utterances were later broken up into words, which could then be recombined to create further utterances.

Mithen favors the holistic approach. As evidence, he points to the fact that all nonhuman primate utterances, such as vervet monkeys' alarm calls, rhythmic chatters of geladas, duets of pair-bonded gibbons, and pant-hoots of chimpanzees, are holistic and indivisible.[8] In other words, nonhuman primates do not have words, even though they have languages and their utterances as a whole convey specific meanings and emotions. Some primatologists disagree, however, and point out, in support of the compositional approach, that Diana and Campbell's monkey calls have both syntactic and semantic rules, which can be used to combine elements ("words") to produce further utterances.[9] The debate on the origin of human language between the compositional and holistic approaches is far from closed.

Humans and Monkeys Can Communicate with Each Other

Here's something very interesting. Studies demonstrate that meanings and emotions of primate utterances may be shared by different primate species. For example, when macaque vocalizations that are made in specific social contexts as expressions of contentment, pleading, dominance, anger, and fear are recorded and then played back, Finnish children and adults are able to interpret accurately what the expressed emotions are.[10] In other words, humans can understand what monkeys mean when they speak!

Another study shows that words spoken by Finnish and English speakers in the social contexts of contentment, pleading, dominance, anger, and fear have the same acoustic waveforms as the macaque vocalizations made in the corresponding contexts.[11] It is as though humans and macaques may be able to communicate with each other through the use of *holistic utterances and messages.*

Mithen contends that human proto-language was *holistic, manipulative* (it was designed to induce desired emotions and behavior in other individuals), *multi-modal* (it involved not only vocal utterances but also gesture and dance), *musical* (the utterances had distinct pitches, rhythms, and melodies), and *mimetic* (conscious and intentional). This proto-language eventually evolved

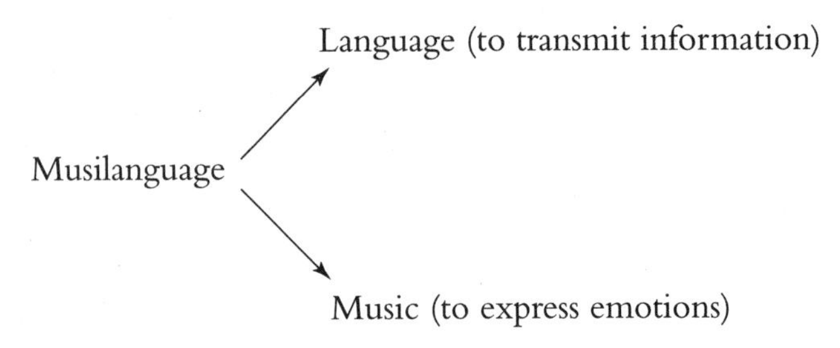

Figure 10.1 Mithen's view on the evolution of music and language

into two systems of communication: *music* to express emotions, and *language* to transmit information.

To demonstrate the common evolutionary origin of music and language, Mithen surveyed a large number of clinical cases of individuals with amusia (the absence of musical abilities while retaining some linguistic abilities) and aphasia (the absence of linguistic abilities while retaining some musical abilities).[12] These case studies largely show that music and language are based on discrete modules in the brain; some of these are separate and dedicated to one or the other *while others are shared.*

Songs Are Evolutionarily Familiar, but Instrumental Music Is Evolutionarily Novel

If Mithen is right, if music and language share a common evolutionary origin in holistic, musical utterances designed to convey messages, one possible implication is that music, in its evolutionary origin, was *songs* that individuals sang to express their desires and emotions, in an attempt to induce desired emotions and behavior in others. In other words, *music in its evolutionary origin was always vocal and never purely instrumental.* Purely instrumental music, unaccompanied by singing, may therefore be evolutionarily novel.

It may be instructive to note in this context that Blackfoot Indians have a word for "song" but not for "instrumental music."[13] The language of the Pirahã in the Amazon forest in Brazil may be an extant example of a musilanguage which Mithen envisions as the precursor to the modern language and music.[14] While the Pirahã language does have words, it has the fewest number of vowels (three) and consonants (seven for women, eight for men) of all known human languages. "The Pirahã people communicate almost as much by singing, whistling, and humming as they do using consonants and vowels. Pirahã prosody is very rich, with a well-documented five-way weight distinction between syllable types."[15]

The former professional musician and current academic linguist, as well as the originator of the holistic approach to the evolutionary origin of language, Alison Wray notes: "To my taste, western classical music (as indeed most other musical traditions worldwide) is different in kind [from musical expressions in evolutionary history]. Its production is, for a start, subject to a heavy burden of learning that few master. There is no naturally facilitated access to the comprehension (let alone creation) of the kinds of melodies, harmonies and rhythms found in the works of Bach or Schoenberg: no equivalent—for music of this kind—of first language acquisition."[16] In other words, according to Wray, classical music of Bach, Schoenberg, and others is *evolutionarily novel*, partly, I contend, because it is largely or entirely instrumental.

Consistent with Wray's assertion, a far greater proportion of the general population can (and spontaneously do) sing songs than play musical instruments. For example, the incidence of tone-deafness in the United Kingdom is estimated to be about 4–5%.[17] In other words, 95% of the population can sing adequately (and some of the tone-deaf people nonetheless often do sing). The proportion of the general population who play musical instruments adequately is nowhere near as high. Further, in many cases of playing musical instruments (such as the guitar or the piano), it is often accompanied by singing.

In the context of the Intelligence Paradox, then, Mithen's theory of the evolutionary origins of music suggests that more intelligent individuals are more likely to appreciate purely instrumental music than less intelligent individuals because such music is evolutionarily novel. In contrast, general intelligence has no effect on the appreciation of vocal music. From this perspective, more intelligent people may appreciate classical music more *because it is largely or entirely instrumental.* And more intelligent people should prefer other types of instrumental music as well.

Intelligence and Tastes for Music

Two large, nationally representative data sets have asked questions about the respondent's musical tastes. One is the GSS, which I explain in the Introduction. The other is the 1986 follow-up to the British Cohort Study (BCS). The BCS is very similar to the NCDS; in fact, it was modeled after the NCDS. Just like the NCDS, the original sample of the BCS includes *all babies* born in Great Britain during one week in April 1970, and they have been followed periodically ever since. In 1986, when the respondents were 16, the BCS asked them a series of questions about what kinds of music they listened to.

In 1993, the GSS asked about the respondents' taste for 18 different kinds of music: "big band," "bluegrass," "country western," "blues or R&B," "Broadway musicals," "classical," "folk," "gospel," "jazz," "Latin," "easy listening," "new age," "opera," "rap," "reggae," "contemporary rock," "oldie," and "heavy metal." Even though it's difficult to classify entire genres of music as either entirely vocal or entirely instrumental, I nevertheless classify big band, classical and easy listening ("elevator music") as largely instrumental, and the rest as largely vocal.

Jazz is a difficult case, as much of jazz is also purely instrumental. However, an earlier study[18] found that more intelligent individuals prefer to listen to jazz. So classifying it as largely vocal, rather than largely instrumental, constitutes a *conservative* classification, which would make it more difficult for me to find evidence for the prediction that more intelligent people are more likely to prefer instrumental music. When it comes to statistical analysis, it is *always* good to be conservative rather than liberal. I believe the fact that so many established musical genres are vocal, and so few are instrumental, is in itself evidence that music in its evolutionary origin might have been vocal.

The 1986 follow-up to the BCS asked about the respondent's taste for 12 different kinds of music: "classical," "light music," "folk music," "disco," "reggae," "soul," "heavy rock," "funk," "electronic," "punk," "other pop music," and "other." I classify classical and light music (easy listening or "elevator music") as largely instrumental, and the rest as largely vocal.

In addition, the 1986 follow-up to the BCS also asked the teenagers' TV viewing habits, by asking whether they watched 22 different types of TV shows. Two of these 22 types refer to music: pop/rock music, and classical music. I therefore examine the effect of intelligence on watching TV programs on different genres of music.

Despite the fact that these surveys were conducted in different decades in different countries, they both support the prediction derived from the Intelligence Paradox.[19] In the US, net of age, race, sex, education, family income, religion, whether currently married, whether ever married, and number of children, more intelligent individuals are more likely to prefer largely instrumental music (big band, classical, and easy listening) than less intelligent individuals. In contrast, intelligence is not associated with the GSS respondents' preference for largely vocal music. Intelligence is also significantly associated with the *difference* in preference between instrumental and vocal music (the mathematical difference obtained by subtracting their average preference for vocal music from their average preference for instrumental music). The more intelligent the GSS respondents are, the greater their difference in preference between the two types of music.

The results are exactly the same for the 1986 BCS sample in the United Kingdom. Net of academic performance (all BCS respondents are still in school), sex, race, religion, family income, mother's education, and father's education, more intelligent British teenagers are more likely to prefer instrumental music (classical and light music) than their less intelligent classmates. In contrast, their intelligence is not associated with their preference for vocal music. As a result, their intelligence is also significantly associated with the *difference* in preference between instrumental and vocal music. The more intelligent the BCS respondents are, the greater the difference in preference between the two types of music.

Net of the same potential confounds, more intelligent BCS respondents are more likely to watch TV shows about classical music than their less intelligent classmates, despite the fact that more intelligent people enjoy watching TV less in general than less intelligent people.[20] In sharp contrast, more intelligent BCS respondents are *less* likely to watch TV shows about pop/rock music. As a result, intelligence is very strongly associated with the difference in the viewing frequency of TV shows about the two types of music.

For illustrative purposes, here is the association between intelligence and preference for classical music among the GSS respondents. The GSS respondents who "like classical music very much" have the average IQ of 106.5. In contrast, those who "dislike classical music very much" (like myself) have the average IQ of 93.3. People who like classical music very much are more intelligent than those who dislike it very much by more than 13 IQ points! As you can see, the association between intelligence and preference for classical music is monotonic; the more they like classical music, the more intelligent they are. The probability

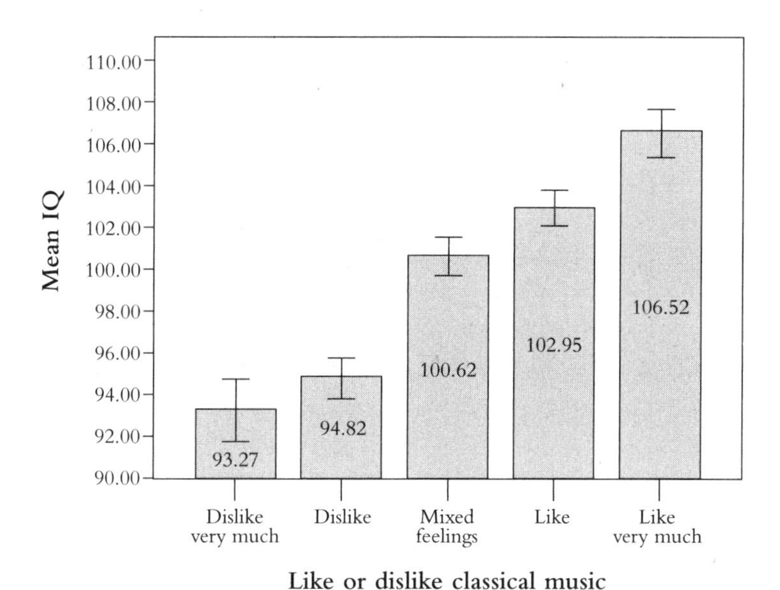

Figure 10.2 Association between intelligence and preference for classical music: GSS

that one would observe a pattern as strong as the one depicted in Figure 10.2 purely by chance, when there is actually no association between verbal intelligence and preference for classical music, is less than one in 100 quadrillion (or 100 thousand trillion or 100 million billion or 10^{17})!

And here is the association between intelligence and whether they usually listen to classical music among the BCS respondents. As you can see, British teenagers who usually listen to classical music are much more intelligent than their classmates who don't usually listen to classical music, by more than 7 IQ points. The probability that one would observe a pattern as strong as the one depicted in Figure 10.3 purely by chance, when there is actually no association between verbal intelligence and preference for classical music, is less than one in 10 nonillion (that's 1 followed by 31 zeroes or 10^{31}). In other words, it's less likely than impossible.

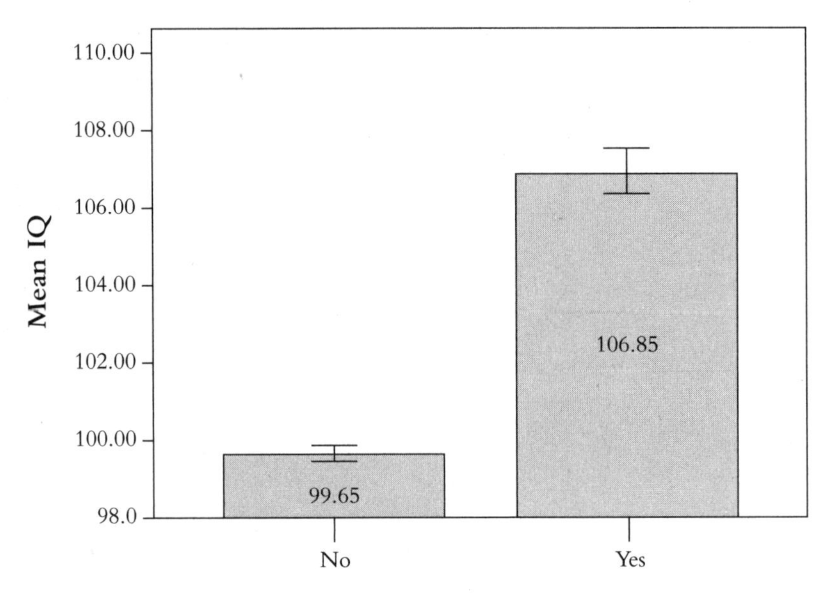

Figure 10.3 Association between intelligence and preference for classical music: BCS86

Evolutionary Novelty or Cognitive Complexity?

The analyses of two separate large, nationally representative data sets, with one sample of teenagers in the United Kingdom and another of adults in the United States, suggest that more intelligent individuals are more likely to prefer evolutionarily novel instrumental music than less intelligent individuals, while intelligence does not affect individuals' preference for evolutionarily familiar vocal music. One potential objection to this conclusion is that the dimension of evolutionary novelty, captured by the distinction between instrumental and vocal music, is confounded with cognitive complexity of music, which is defined by chordal complexity (the number of chords, tones, and instruments used in the music and their interrelationships).

For example, classical music, which is largely instrumental, is also cognitively complex. It is probably the most cognitively complex form of music in human history. On the other extreme, rap music, which is almost entirely vocal, often to the exclusion of any discernible melodic structure, is also cognitively very simple. So critics may argue that the association between intelligence and preference for instrumental music may really be an association between intelligence and cognitively complex forms of music.

In order properly to examine and rule out this alternative hypothesis, I would ideally need a quantitative "cognitive complexity score" for each genre of music, in the form (classical = 5, jazz = 4.5, etc.). Further, such "cognitive complexity scores" would ideally have been validated and widely in use. I searched the literature extensively, and consulted several different experts in music perception, but have not been able to locate such "cognitive complexity scores" for different musical genres. They simply do not seem to exist. Yet most people seem to understand and agree intuitively that, for example, classical music and jazz are far more cognitively complex than, say, rap music. In the absence of quantitative and validated "cognitive complexity scores," I must rely on such intuitive but widely shared senses of cognitive complexity of musical genres.

A potential problem with inspecting the association between intelligence and preference for each specific musical genre is that preferences for all musical genres are very highly correlated. It appears that there are people who like music and there are those (like me) who don't, and those who like music like all types of music. For example, in the GSS data, preference for classical music is positively and significantly correlated with preference for both bluegrass and reggae music. In fact, it is significantly positively associated with 12 out of the 17 other genres of music. People who like classical music like to listen to most genres of music.

In Chapter 3, I explain the statistical technique called factor analysis. This is the technique psychometricians use to extract a common underlying factor—general intelligence—that explains individuals' performance on all types of cognitive tests. Factor analysis can be used with any set of scores to extract what's common among them. If you submit GSS respondents' preference for 18 different genres of music, factor analysis extracts only one underlying dimension. It means that one dimension—general preference for music—explains individuals' preference for *all* types of music, just as one dimension—general intelligence—explains individuals' performance on all types of cognitive tests.

So I examine the association between intelligence and preference for each genre of music, while holding constant preferences for all other types of music. The result shows that, as expected, preference for classical music is significantly positively associated with intelligence, net of preferences for all other types of music. However, preference for big band is even more strongly positively associated with intelligence than is preference for classical music. It would be difficult to make the case that big band music is more cognitively complex than classical music.

At the other extreme, as suspected, preference for rap music is significantly negatively associated with intelligence. However, preference for gospel music is even more strongly negatively associated with intelligence. It would be difficult to make the case that gospel is less cognitively complex than rap. (I might also point out in passing that, with its close link to religious rituals, gospel is a particularly evolutionarily familiar form of music.[21])

At the same time, preference for opera, another highly cognitively complex form of music, is not significantly correlated with intelligence. Its nonsignificantly positive association is smaller than those for oldies, reggae, and Broadway musicals. It would be difficult to make the case that oldies, reggae, and Broadway musicals are cognitively more complex than opera.

These conclusions remain when I further control for the GSS respondent's age, race, sex, education, family income, religion, whether currently married, whether ever married, and number of children. When these additional controls are included in the statistical analysis, the positive association between preference for classical music and intelligence is no longer statistically significant, while the association between preference for big band and intelligence remains statistically significantly positive. With the additional controls, the association between preference for oldies and intelligence is statistically significantly positive. It would be difficult to make the case that oldies and big band are cognitively more complex than classical music. Other researchers[22] have classified blues, jazz, classical and folk music as "structurally complex," but, when preferences of all musical genres are controlled, preferences for none of these "structurally complex" genres are significantly correlated with intelligence. Preferences for folk and jazz are nonsignificantly *negatively* associated with intelligence. All in all, the analysis provides very little support for the view that more intelligent individuals necessarily and uniformly prefer cognitively complex genres of music.

The Intelligence Paradox applied to musical genres might therefore explain, among other things, why Jews and East Asians are more likely to excel as classical musicians and why most NPR stations throughout the country play classical and jazz music as their station themes. Purely instrumental music may be evolutionarily novel and therefore unnatural, and more intelligent individuals may have a preference for such unnatural genres of music.

Chapter 11

Why Intelligent People Drink and Smoke More

Recall from the Introduction that one of my major goals in this book is to break the equation of intelligence with human worth and to demolish the myth that intelligence is universally good and that more intelligent people are universally better at everything than less intelligent people. For this purpose, an interesting test case involves the consumption of alcohol, tobacco, and drugs, because it is widely agreed that their consumption—especially their excessive use—has largely negative health consequences. If I can demonstrate that more intelligent individuals are more likely to consume these substances to excess, then I would have gone a long way toward demonstrating that more intelligent people don't always do the right thing and, in fact, more intelligent people are often more likely to do stupid things.

In this chapter, I'm therefore going to discuss drinking alcohol, smoking cigarettes, and using illegal drugs. While, once again, science does not make any value judgment, it would be very difficult to make the case from any perspective other than strictly scientific that these activities are neither good nor bad. We know the health hazards of drinking, smoking, and taking drugs. Although the current consensus of medical researchers appears to be that drinking alcohol *in moderation* may have some health benefits, you will see below that that's not what I am talking about. I will be talking about getting drunk and binge drinking, which have no discernible health benefits.

So, unlike all the preferences and values I have discussed so far in this book, drinking alcohol, smoking cigarettes, and taking drugs, especially to excess, are all inherently bad for health. But the Intelligence Paradox is not about good or bad (healthy or unhealthy) values and preferences. All that matters is their evolutionary novelty. No matter how bad they may be, the Intelligence Paradox would nevertheless predict that more intelligent individuals are more likely to value and prefer them *if* they are evolutionarily novel. So the question is, are they?

Brief Histories of Alcohol, Tobacco, and Drugs

Alcohol

The human consumption of alcohol probably originated from *frugivory*—consumption of fruits.[1] Fermentation of sugars by yeast naturally present in overripe and decaying fruits produces ethanol, known to intoxicate birds and mammals that consume them.[2] However, the amount of ethanol alcohol present in such fruits ranges from trace to 5%, roughly comparable to light beer (0–4%). It is nothing compared to the amount of alcohol present in regular beer (4–6%), wine (12–15%), and distilled spirits (20–95%).

"Ingestion of alcohol, however, was unintentional or haphazard for humans until some 10,000 years ago,"[3] and "intentional fermentation of fruits and grain to yield ethanol arose only recently within human history."[4] The production of beer, which relies on a large amount of grain, and that of wine, which similarly require a large amount of grapes, could not have taken place before the advent of agriculture around 8,000 BC. Archeological evidence dates the production of beer and wine to Mesopotamia at about 6,000 BC.[5] The origin of distilled spirits is far more recent, and is traced either to the Middle East or China at about 700 AD. The word *alcohol*—*al kohl*—is Arabic in origin.

"Relative to the geographical duration of the hominid lineage, therefore, exposure of humans to concentration of ethanol higher than those attained by fermentation alone [i.e., at most 5%] is strikingly recent."[6] Further, any "unintentional or haphazard" consumption of alcohol in the ancestral environment, via the consumption of overripe and decaying fruits, happened as a result of *eating,* not *drinking,* whereas alcohol is almost entirely consumed today via drinking. So there appears very little doubt that drinking alcohol of any measureable concentration is evolutionarily novel.

Tobacco

The human consumption of tobacco is more recent in origin than that of alcohol. The tobacco plant originated in South America and spread to the rest of the world.[7] Native Americans began cultivating two species of the tobacco plant (*Nicotiana rustica* and *Nicotiana tabacum*) about 8,000 years ago.[8] The consumption of tobacco was unknown outside of the Americas until Columbus brought it back to Europe at the end of the 15th century.[9] The consumption of tobacco is therefore of very recent historical origin and is definitely evolutionarily novel.

Drugs

Most psychoactive drugs have even more recent historical origin than alcohol and tobacco. "Before the rise of agriculture, access to psychoactive substances likely was limited."[10] The use of opium dates back to about 5,000 years ago,[11] and the earliest reference to the pharmacological use of cannabis is in a book written in 2737 BC by the Chinese Emperor Shen Nung.[12] Most other psychoactive drugs commonly in use today require modern chemical procedures to manufacture, and are therefore of much more recent origin. Morphine was isolated from opium in 1806.[13] Heroin was discovered in 1874.[14] Cocaine was first manufactured in 1860.[15] It is therefore safe to conclude that most psychoactive drugs in use today are evolutionarily novel and very recent in historical origin.

Given that the consumption of alcohol, tobacco, and psychoactive drugs is all evolutionarily novel—unknown before the end of the Pleistocene Epoch 10,000 years ago—the Intelligence Paradox would predict that, perhaps contrary to common sense and the unquestioned assumption that intelligent people make "smart" choices in life, more intelligent individuals are more likely to consume all such substances than less intelligent individuals. Both NCDS and Add Health allow me to examine the effect of childhood intelligence on adult substance use.

Intelligence and Substance Use

Alcohol

NCDS asks its respondents about the *frequency* and *quantity* of their alcohol consumption. First, it asks its respondents how often they usually have an alcoholic drink at ages 23, 33, and 42. I use factor analysis to compute the NCDS respondents' latent tendency to consume alcohol frequently in their adult life. (Latent factors produced by factor analysis have the mean of 0 and standard deviation of 1.)

Then NCDS asks its respondents about the quantity of their consumption of different alcoholic beverages such as beer, spirits, wine, martini, sherry, and "alcopops" (that's British for flavored alcoholic drinks like wine coolers). NCDS asks these questions at ages 23, 33, and 42. Once again, I use factor analysis to compute the NCDS respondents' latent tendency to consume a large quantity of various alcoholic beverages.

The analysis of the NCDS data shows that, net of sex, religion, religiosity, whether currently married, whether ever married, number of children, education, income, whether depressed, satisfaction with life, parents' social class, mother's education, and father's education, more intelligent children are more likely to grow up to consume more alcohol in their adult life, measured both by frequency and quantity. Because such a large number of potential confounds are statistically controlled for, it is not likely (albeit technically possible) that the association between childhood general intelligence and adult alcohol consumption can be attributed to another factor.

For example, it's not likely that it is because more intelligent people are more likely to have certain occupations—such as executive or managerial positions that require socialization or negotiation over drinks—that childhood general intelligence and adult alcohol consumption are positively associated, because both education and income, as well as social class at birth, mother's education, and father's education, are controlled for. Interestingly, of these factors, only income and father's education independently increase the respondent's adult alcohol consumption, both by frequency and quantity. Education, social class at birth, and mother's education have no effect on adult alcohol consumption.

Add Health asks four questions about their alcohol consumption: How much they drink, how often they drink, how often they engage in binge drinking (five or more drinks in one sitting), and how often they get drunk. Once again, using factor analysis, I compute Add Health respondents' latent tendency to consume alcohol.

The analysis of the Add Health data shows that, net of age, sex, race, Hispanicity, religion, marital status, parenthood, education, income, political attitude (liberal vs. conservative), religiosity, general satisfaction with life, whether they are taking medication for stress, whether they feel stress but do not take medication for it, frequency of socialization with friends, number of sex partners in the last 12 months, childhood family income, mother's education, and father's education (in other words, *lots* of potentially confounding factors), childhood intelligence significantly increases adult alcohol consumption. The more intelligent they are in junior high and high school, the more alcohol they consume in early adulthood.

Once again, given the even larger number of statistical controls in the analysis of the Add Health data than in the analysis of the NCDS data, it is very unlikely that the apparent effect of childhood intelligence can be attributed to something else. Neither income nor education is significantly associated with adult alcohol consumption. As with NCDS, father's education (but not mother's education) increases Add Health respondents' adult alcohol consumption.

Figures 11.1 and 11.2 show the association between childhood general intelligence (grouped into "cognitive classes") and adult alcohol consumption, by frequency and by quantity, from the NCDS data. As you can see, the association is perfectly monotonic. The more intelligent NCDS respondents are in childhood, the more they consume alcohol in adulthood. In Figure 11.1 for frequency, "very bright" individuals and "very dull" individuals

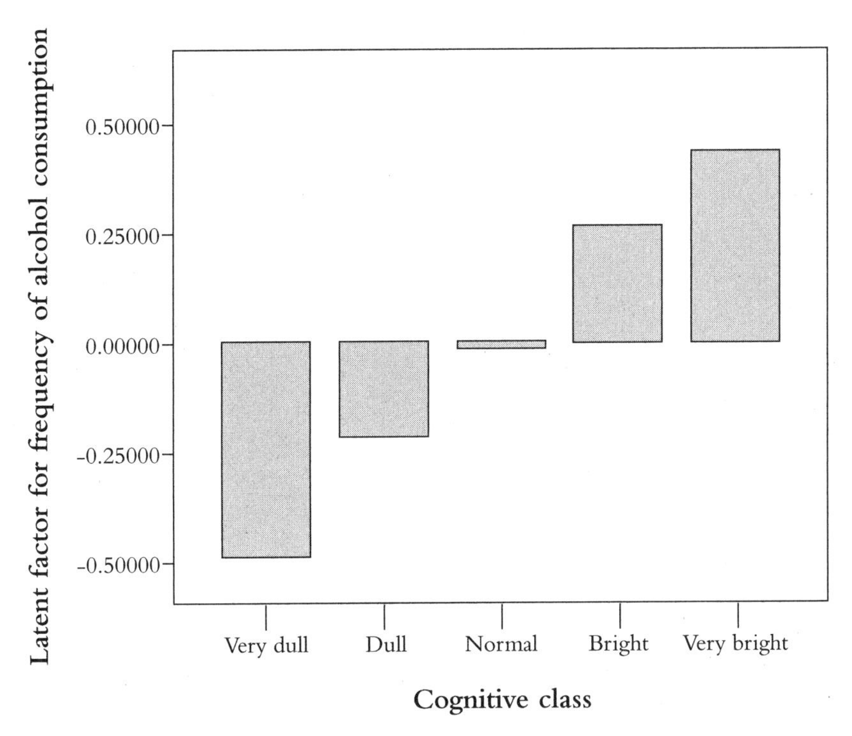

Figure 11.1 Association between childhood general intelligence and frequency of alcohol consumption (NCDS)

are separated by nearly a full standard deviation. In Figure 11.2 for quantity, they are separated by four-fifths of a standard deviation. These effects are very large.

More Intelligent People Are More Likely to Binge Drink and Get Drunk

There are occasional medical reports and scientific studies which tout the health benefits of *mild* alcohol consumption, such as drinking a glass of red wine with dinner every night. So it may be tempting to conclude that more intelligent individuals

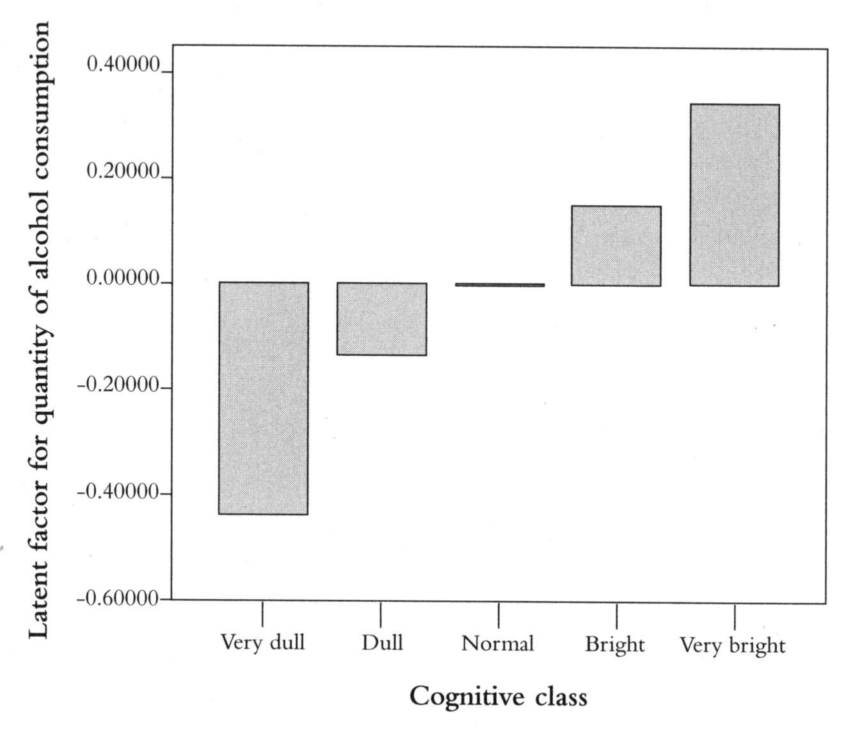

Figure 11.2 Association between childhood general intelligence and quantity of alcohol consumption (NCDS)

are more likely to engage in such mild alcohol consumption than less intelligent individuals, and the positive association between childhood general intelligence and adult alcohol consumption reflects such mild, and thus healthy and beneficial, alcohol consumption.

Unfortunately for the intelligent individuals, this is not the case. More intelligent children are more likely to grow up to engage in *binge drinking* (consuming five or more units of alcohol in one sitting) and *getting drunk*.

Add Health asks its respondents specific questions about binge drinking and getting drunk. For binge drinking, Add Health asks: "During the past 12 months, on how many days did you drink

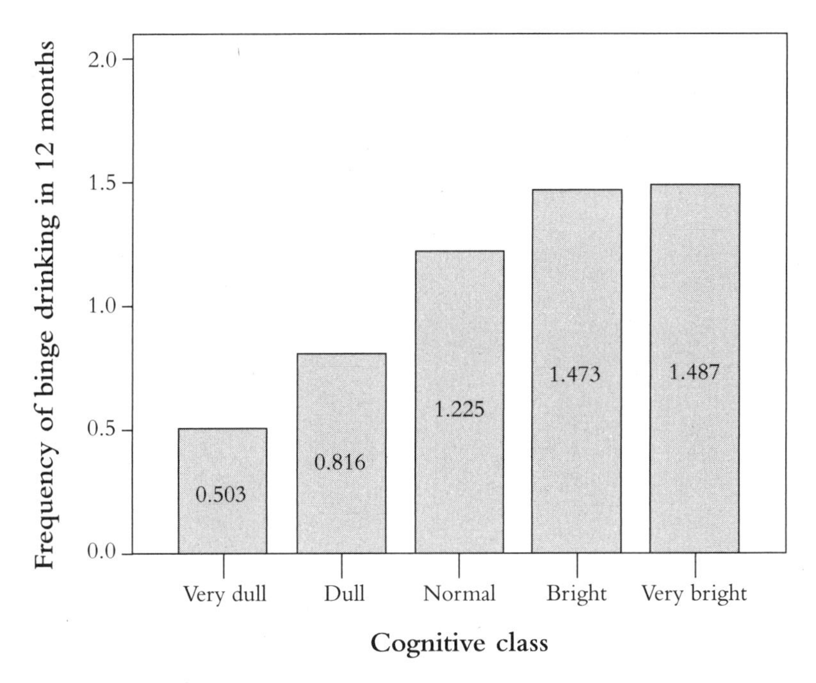

Figure 11.3 Association between childhood general intelligence and frequency of binge drinking (Add Health)

five or more drinks in a row?" For getting drunk, it asks: "During the past 12 months, on how many days have you been drunk or very high on alcohol?" For both questions, the respondents can answer on a six-point ordinal scale: 0 = none, 1 = 1 or 2 days in the past 12 months, 2 = once a month or less (3 to 12 times in the past 12 months), 3 = 2 or 3 days a month, 4 = 1 or 2 days a week, 5 = 3 or 5 days a week, 6 = every day or almost every day.

As you can see in Figure 11.3, there is a clear monotonic positive association between childhood intelligence and adult frequency of binge drinking. "Very dull" Add Health respondents (with childhood IQ < 75) engage in binge drinking less than once a year. In sharp contrast, "very bright" Add Health respondents (with childhood IQ > 125) engage in binge drinking roughly once every other month.

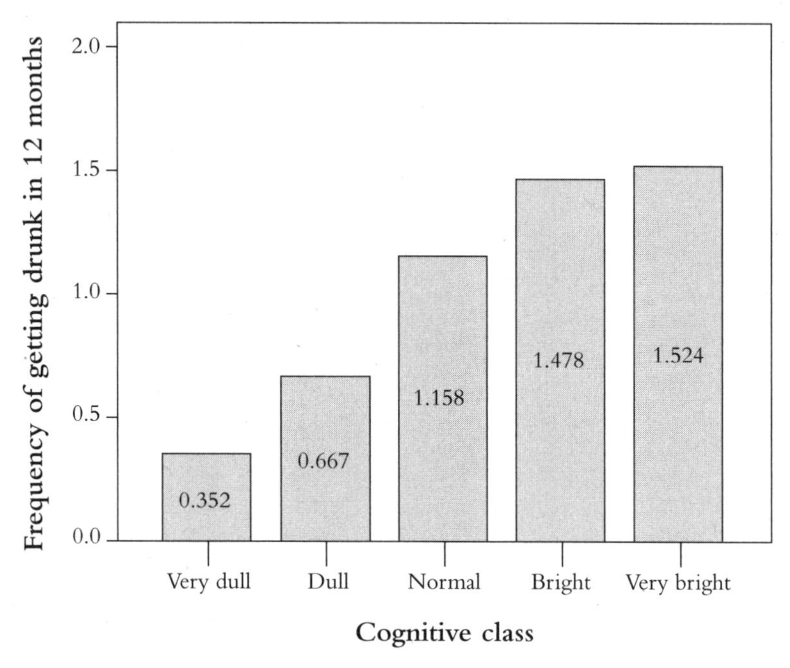

Figure 11.4 Association between childhood general intelligence and frequency of getting drunk (Add Health)

The association between childhood intelligence and adult frequency of getting drunk is equally clear and monotonic, as you can see in Figure 11.4. "Very dull" Add Health respondents almost never get drunk, whereas "very bright" Add Health respondents get drunk once every other month or so, just as frequently as they engage in binge drinking, which makes sense, since binge drinking almost necessarily and by definition would make most people drunk.

In a multiple ordinal regression, childhood intelligence has a significantly positive effect on adult frequency of both binge drinking and of getting drunk ($ps < .00001$), controlling for age, sex, race, ethnicity, religion, marital status, parental status, education, earnings, political attitudes, religiosity, general satisfaction with life, taking medication for stress, experience of stress without taking medication, frequency of socialization with friends, number of sex partners in the last 12 months, childhood family income, mother's education, and father's education. I honestly cannot think of any other variable that might be correlated with childhood intelligence than those already controlled for in the multiple regression analyses. It means that the effect of childhood intelligence is not confounded with any of the variables already included in the equations. It is very likely that childhood intelligence itself, not anything else that may be confounded with it, increases the adult frequency of binge drinking and getting drunk.

Note that education is controlled for in the ordinal multiple regression analysis. Given that Add Health respondents in Wave III (when their drinking behavior is measured) are in their early 20s, it may be tempting to conclude that the association between childhood intelligence and adult frequency of binge drinking and getting drunk may be mediated by college attendance. More intelligent children are more likely to go to college, and college students are more likely to engage in binge drinking and to get drunk. The significant partial effect of childhood intelligence on the adult frequency of binge drinking and getting drunk, net of education, shows that this indeed is *not* the case. Childhood intelligence itself, not education, increases the adult frequency of binge drinking and getting drunk.

In fact, in both equations, education does *not* have a significant effect on binge drinking and getting drunk. Net of all the other variables included in the ordinal multiple regression equations, education is *not* significantly associated with the frequency of binge drinking and getting drunk. Among other things, it means that college students are more likely to engage in binge drinking, not because they are in college, but because they are more intelligent.

Tobacco

NCDS measures its respondents' tobacco consumption by asking how many cigarettes a day they usually smoke at ages 23, 33, 42, and 47. I compute their latent tendency toward tobacco consumption by performing factor analysis.

To my surprise, and contrary to the prediction of the Intelligence Paradox, the analysis of the NCDS data shows that, net of the same control variables as above in the analysis of alcohol consumption, more intelligent British children are *less* likely to grow up to consume tobacco in their adult life.

Add Health measures its respondents' tobacco consumption by asking on how many days they have smoked cigarettes in the last 30 days and how many cigarettes a day they smoked in the last 30 days. I compute their latent tendency toward tobacco consumption by performing factor analysis.

In sharp contrast to NCDS, and consistent with the prediction of the Intelligence Paradox, Add Health data confirm the prediction. Net of the same control variables as before in the analysis of alcohol consumption, more intelligent children grow up to consume more tobacco in their early adulthood.

The divergent effect of childhood intelligence on adult tobacco consumption is clear in Figures 11.5 and 11.6. Figure 11.5 depicts the monotonically negative association between childhood intelligence and adult tobacco consumption among the NCDS respondents in the United Kingdom. The more intelligent they are before the age of 16, the less tobacco they consume in their 20s, 30s, and 40s. Figure 11.6 shows the largely positive association between intelligence and adult tobacco consumption among the Add Health respondents in the United States. The association is not monotonically positive, but still "normal," "bright," and "very bright" Add Health respondents consume more tobacco than their "very dull" and "dull" counterparts.

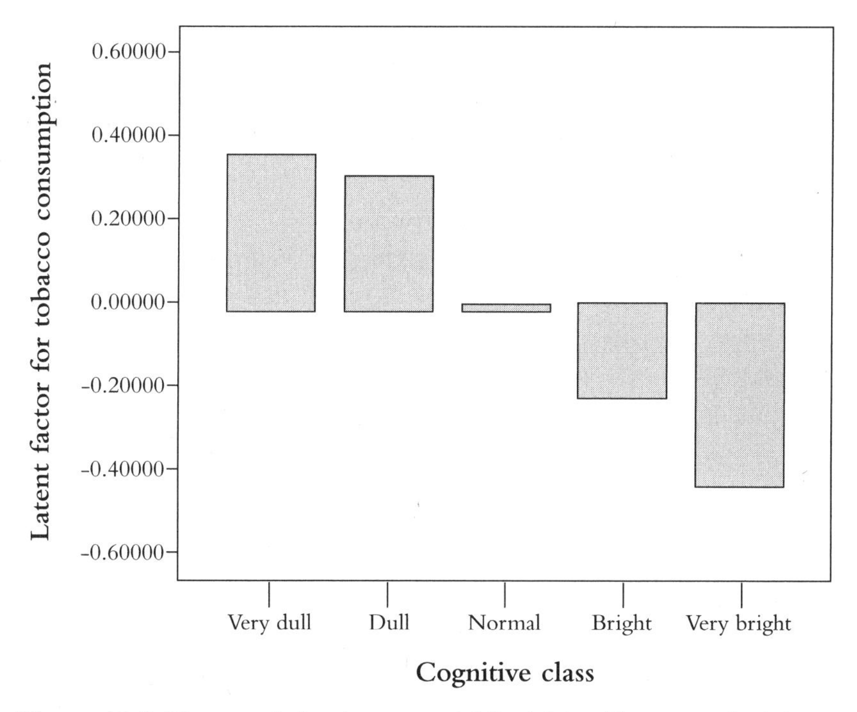

Figure 11.5 The association between childhood intelligence and adult tobacco consumption (NCDS)

Why Does Intelligence Affect Smoking Differently in the US and the UK?

I'm not sure what accounts for the divergent results from NCDS and Add Health when it comes to the effect of childhood intelligence on adult smoking. However, mine is not the only study which shows such varied results. Other studies[16] have also shown that more intelligent Brits are less likely to smoke, while more intelligent Americans are more likely to smoke.

In my study, the two data sets are different in two major respects. First, NCDS is conducted in the United Kingdom, while Add Health is conducted in the United States. Second, NCDS respondents were born in March 1958, while Add Health

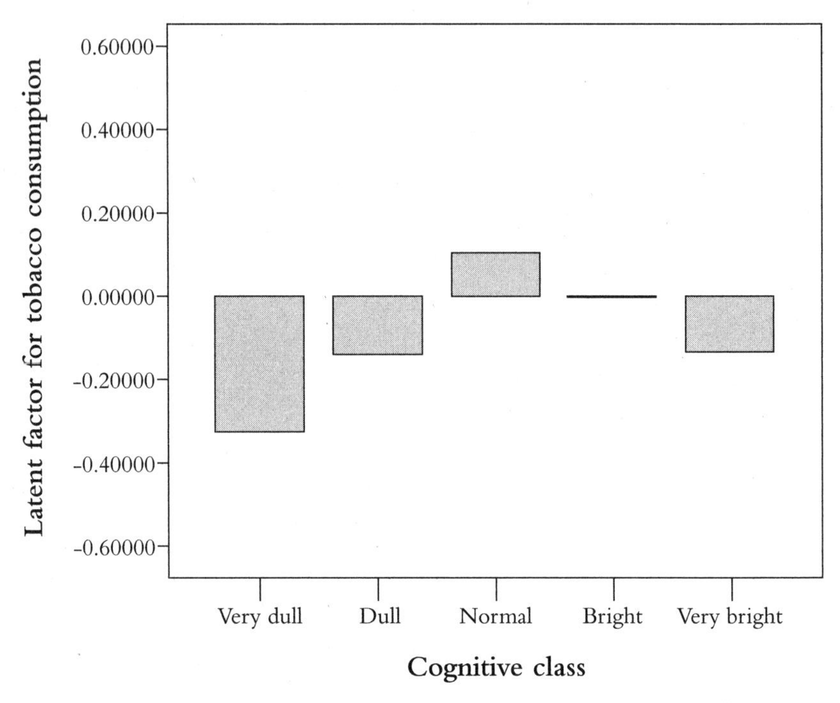

Figure 11.6 The association between childhood intelligence and adult tobacco consumption (Add Health)

respondents were born between 1974 and 1983. Further research is necessary to determine whether it is the cultural differences between the two (otherwise very similar) nations or the generational differences between the NCDS and Add Health cohorts that produce the strikingly divergent results when it comes to the effect of childhood intelligence on adult tobacco consumption.

Among the possible differences between the US and the UK, the public anti-smoking campaign has been far more aggressive and blatant in the UK than in the US. For example, in the US, each pack of cigarettes carries the Surgeon General's (relatively tame and clinical) warning ("Smoking causes lung cancer, heart disease, emphysema, and may complicate pregnancy") *in small print, on the side of the package.* In the UK, the warnings are much more blatant and graphic ("Smoking kills," "Smokers die younger," "Smoking may reduce the blood flow and causes impotence," "Smoking can cause slow and painful death," "Smoking clogs the arteries and causes heart attacks and strokes," "Smoking when pregnant harms your baby") *in extremely large print, on the front of the package.* Note that death is never mentioned explicitly in the Surgeon General's warning in the US, but is frequently mentioned in the UK warnings. And they mention something much worse than death from an evolutionary perspective: impotence (for men) and harm to the baby (for women), and thereby implied lack of reproductive success.

When I saw the warning "Smoking kills" for the first time in 2003, on a pack of cigarettes that my LSE colleague was smoking, I thought it was a joke. It looked like a gag item that one might buy at a novelty store in a shopping mall, like Spencer's or Hot Topic. I didn't realize that it was for real until after I saw other packs of cigarettes with similar warnings later.

Because government warnings and public campaigns (as well as the written language as their medium of communication) are themselves evolutionarily novel, more intelligent individuals may be more likely to respond to such warnings than do less intelligent individuals. This is just one of the possible reasons why intelligence may have such starkly opposite effects on smoking in the US and the UK.

To be honest, I don't find this a particularly convincing answer myself. I feel relatively certain that the national difference in the effect of general intelligence on smoking is robust and real, not a methodological artifact, because different studies using different data sets and methodologies all confirm it. But I don't like my own explanation for it. I feel there is a better explanation out there; I just don't know what it is.

Drugs

At age 42 only, NCDS asks its respondents whether they have ever tried 13 different types of illegal psychoactive drugs: cannabis, ecstasy, amphetamines, LSD, amyl nitrate, magic mushrooms, cocaine, temazepan, semeron, ketamine, crack, heroine, and methadone. Via factor analysis, I compute NCDS respondents' latent tendency to consume illegal drugs.

The statistical analysis of the NCDS data shows that, net of the same control variables as before, more intelligent children are more likely to grow up to consume more illegal drugs than less intelligent children. The higher their general intelligence before the age of 16, the more illegal drugs that they try before the age of 42.

Figure 11.7 shows the association between childhood intelligence, grouped by "cognitive class," and latent adult tendency to consume illegal drugs. Just as with alcohol consumption, there is a monotonic positive association between childhood general intelligence and adult consumption of illegal drugs. But the effect of childhood general intelligence on adult consumption of illegal drugs is not as large as its effect on adult alcohol consumption. "Very bright" and "very dull" NCDS respondents are separated only by about one third of a standard deviation.

Add Health asks its respondents about their consumption of five different illegal drugs: marijuana, cocaine, LSD, crystal meth, and heroine. Via factor analysis, I once again compute Add Health respondents' latent tendency to consume illegal drugs.

The statistical analysis of the Add Health data shows that the effect of childhood intelligence on adult consumption of illegal drugs, while positive as predicted by the Intelligence Paradox, is not statistically significant. So Add Health data do not provide unambiguous support for the prediction of the Intelligence Paradox with regard to illegal drugs as do the NCDS data.

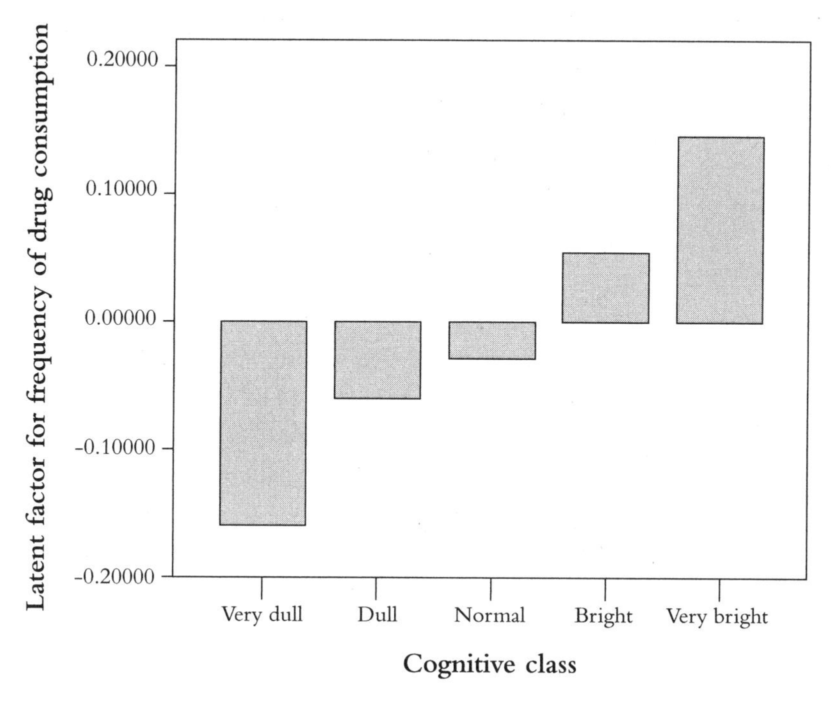

Figure 11.7 Association between childhood general intelligence and frequency of illegal drug consumption

Intelligence and Criminality

Criminologists have long known that criminals on average have lower intelligence than the general population.[17] Juvenile delinquents are less intelligent than nondelinquents,[18] and a significant difference in IQ between delinquents and nondelinquents appears as early as ages 8 and 9.[19] Chronic offenders are less intelligent than one-time offenders,[20] and serious offenders are less intelligent than less serious offenders.[21] The negative association between general intelligence and criminality is not an artifact of a selection bias, whereby less intelligent criminals are more likely to be caught than more intelligent criminals who get away, because the association exists even in self-report studies that do not rely on official police statistics.[22] (Yes, people often do reveal in interviews and surveys that they have committed crimes that they are not charged with, even those that the police don't know about.)

Why is this? Why do criminals have lower intelligence than the general population? And why do more chronic and serious criminals have lower intelligence than their less chronic and serious counterparts?

From the perspective of the Intelligence Paradox, there are two important points to note. First, much of what we now call interpersonal crime today, such as murder, assault, robbery, and theft, were probably routine means of competition among men for resources and mates. This is how men likely competed for resources and mating opportunities for much of human evolutionary history. We may infer this from the fact that behaviors that would be classified as criminal if engaged in by humans are common among other species,[23] including other primates such as chimpanzees,[24] bonobos,[25] and capuchin monkeys.[26]

Second, the institutions and technologies that control, detect, and punish criminal behavior in society today—the police, the courts, the prisons, CCTV cameras, DNA fingerprinting—are all evolutionarily novel. There was very little formal third-party enforcement of norms in the ancestral environment, only second-party enforcement (retaliation by victims and their kin and allies) or informal third-party enforcement (ostracism). (One need only recall van Beest and Williams's findings in their Cyberball experiment, discussed in Chapter 2, to realize how powerful a punishment ostracism must have been in the ancestral environment.)

It therefore makes sense from the perspective of the Intelligence Paradox that men with low general intelligence may be more likely to resort to evolutionarily familiar means of competition for resources (theft rather than full-time employment) and mating opportunities (rape rather than computer dating), because they are less likely to recognize and comprehend the more evolutionarily novel means (as well as often lacking access to them). It also makes sense that such men do not fully comprehend the consequences of criminal behavior imposed by evolutionarily novel technologies and institutions of law enforcement.

Why Do Less Intelligent People Commit Some Crimes but Not Others?

But if less intelligent individuals are more likely to commit crimes, why are they less likely to use illegal drugs? The consumption of psychoactive drugs, such as marijuana, cocaine, and heroin, is illegal in both the UK and the US; in other words, it is a crime to use such substances. So why aren't less intelligent individuals more likely to commit the crime of drug use?

This is the proverbial case of "the exception which proves the rule." As I mention above, less intelligent individuals are more likely to engage in crime, not because it is criminal per se, but because most of it is evolutionarily familiar. Less intelligent individuals are less likely to engage in behavior that is evolutionarily novel, whether it is defined by the civilized society as criminal or not.

This is why less intelligent individuals are less likely to consume evolutionarily novel substances of psychoactive drugs, even though it is criminal. Less intelligent individuals are probably less likely to engage in other evolutionarily novel forms of criminal behavior, such as check forgery, insider trading, and embezzlement, even apart from the fact that more intelligent individuals are probably more likely to have the opportunity to commit such crimes. (That is true for insider trading and embezzlement, but probably not check forgery.)

Less intelligent individuals are simultaneously less likely to consume the legal substance of alcohol and the illegal substances of marijuana, cocaine, and heroin. At the same time, they are more likely to commit the crimes of murder, rape, and theft but not the crime of drug consumption. These two observations suggest that what matters is not legality or criminality per se but evolutionary novelty. Less intelligent individuals *are* more likely to commit crimes, but *only* when they are evolutionarily familiar.

Unlike the implications of the Intelligence Paradox discussed in Chapters 5 through 10, the empirical support for the implication of the Intelligence Paradox with regard to alcohol, tobacco, and illegal drugs is somewhat equivocal. Both NCDS and Add Health data provide strong empirical support for the Intelligence Paradox with regard to alcohol consumption. Both in the UK and in the US, more intelligent children grow up to consume more alcohol than less intelligent children. In addition, more intelligent American children grow up to binge drink and get drunk more frequently in their early adulthood.

The empirical support is split when it comes to tobacco consumption. More intelligent children grow up to consume less tobacco in adulthood in the UK, while the opposite is the case in the US. Only the Add Health data provide support for the prediction of the Intelligence Paradox with regard to tobacco consumption. Finally, NCDS data provide strong empirical support for the implication of the Intelligence Paradox with regard to illegal drugs while Add Health data don't. More research is necessary to figure out what is behind the divergent results for tobacco and illegal drugs in the UK and in the US.

Chapter 12

Why Intelligent People Are the Ultimate Losers in Life

Reproduction Is the Ultimate Goal of All Living Organisms

If any value is deeply evolutionarily familiar, it is reproductive success. If any value is truly unnatural, if there is one thing that humans (and all other species in nature) are decisively *not* designed for, it is voluntary childlessness. All living organisms in nature, including humans, are evolutionarily designed to reproduce. Reproductive success is the ultimate end of all biological existence. While having children is not the only means to achieve reproductive success (representation of one's genes in the next generation), as it could also be achieved by investment in close genetic relatives, it is nonetheless the primary means of maximizing reproductive success. None of us are descended from ancestors who remained childless, and we are disproportionately descended from individuals who achieved disproportionate reproductive success. So voluntary childlessness is not part of evolved human nature, just as exclusive homosexuality is not part of evolved human nature.

In Chapter 9, I address the question of why some people identify themselves as homosexual and engage in homosexual behavior despite the fact that human nature is largely heterosexual. In this chapter, I address the question of why some individuals choose to remain childless or have fewer children than they can safely raise to sexual maturity despite the fact that reproductive success is the ultimate meaning of life.

Having children, and having as many children as one can potentially raise to sexual maturity so that the children themselves can reproduce, is an evolutionarily familiar goal. In contrast, voluntary childlessness, or having far fewer children than one can reasonably raise to sexual maturity, is evolutionarily novel. The Intelligence Paradox would therefore predict that more intelligent individuals are more likely to have fewer children or to remain childless than less intelligent individuals.

Intelligence and the Value for Children

At age 23, near the beginning of the reproductive careers of most British people, NCDS asked its respondents how many children they wanted to have in their lives. More intelligent NCDS respondents wanted significantly fewer children than their less intelligent counterparts. Childhood general intelligence has a significantly negative effect on how many children they want, for both men and women. The more intelligent they are in childhood, the fewer children they want in adulthood.

As you can see in Figure 12.1, women who do not want to have any children at all in their lives have significantly higher childhood IQ than those who want at least one child. Those who

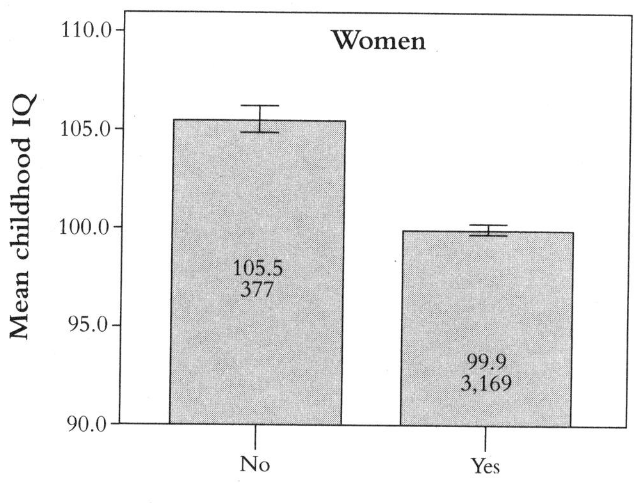

Figure 12.1 Association between mean childhood IQ and desire for parenthood at 23: women (NCDS)

want no children have a mean childhood IQ of 105.5 whereas those who want some children have a mean childhood IQ of 99.9.

The picture is identical among men. Those who do not want to have any children at all have a mean childhood IQ of 104.3 whereas those who want some children have a mean childhood IQ of 100.0. The differences between the two categories of NCDS respondents are highly statistically significant among both women and men.

However, once I control for whether currently married, whether ever married, religiosity, religion, income, education, social class at birth, mother's education, father's education, and number of siblings, childhood IQ has a significantly negative effect on the desired number of children only among men, not among women. Among men, childhood general intelligence still has a significantly negative effect on the desired number of

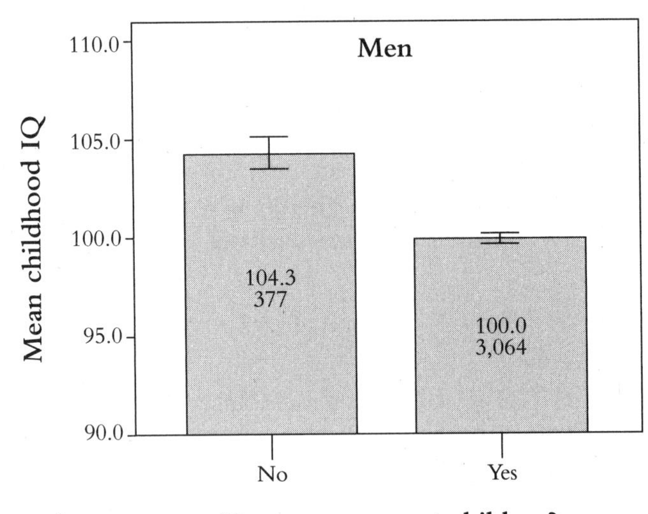

Figure 12.2 Association between mean childhood IQ and desire for parenthood at 23: men (NCDS)

children at age 23. No other variables have any significant effect on the number of desired children for men.

Among women, the number of siblings has a significantly positive effect on the desired number of children. The more siblings the women have themselves, the more children they want to have. But childhood general intelligence no longer has a significant association with the desired number of children once all the social and demographic factors are statistically controlled.

However, even net of the same social and demographic factors, childhood general intelligence has a significantly negative effect on the desired *parenthood*—whether they want to become parents or remain childless—both among men and women. More intelligent men and women are significantly more likely to want to remain childless than less intelligent men and women.

So it appears that general intelligence makes a difference only in the decision to become parents or not for both sexes. Less intelligent individuals are significantly more likely to want to become parents, and more intelligent individuals are significantly more likely to want to remain voluntarily childless. But, beyond that, only less intelligent men, but not less intelligent women, want to have more children than their more intelligent counterparts. Childhood general intelligence significantly predicts the value for *parenthood* for both men and women, but the value for *children* only for men.

Intelligence and the Number of Children

According to one study of the Swedish population, 99.7% of women and 96.5% of men complete their lifetime reproduction by the time they are 45.[1] In other words, very few people—women or men—have children after they are 45. By Sweep 7 in 2004–2005, the NCDS respondents were 46–47 years old, so I can safely assume that most of them have completed their reproductive careers by Sweep 7. I will therefore look at how many children they have actually had before Sweep 7.

By the time they are 47, childhood general intelligence has a significantly negative effect on the number of children NCDS respondents have had only among women, not among men. More intelligent women have had fewer children than less intelligent women, but more intelligent men have not had fewer children than less intelligent men. Recall that, in a bivariate analysis, both more intelligent men and more intelligent women *wanted* to have fewer children when they were 23. It therefore means that more intelligent women have been able to fulfill their desire to have fewer children a quarter of a century later, at the end of their reproductive careers, but more intelligent men have not been able to fulfill their similar desire to have fewer children.

Even net of the same social and demographic characteristics as before, in the analysis of the desired number of children above, more intelligent women have fewer children in their lifetimes than

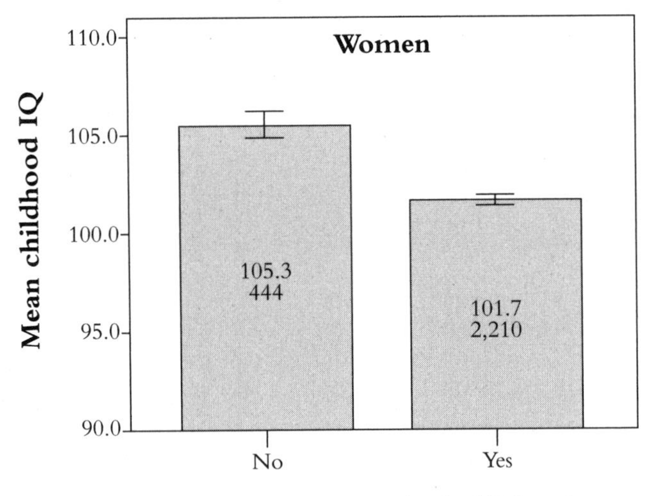

Figure 12.3 Association between mean childhood IQ and parenthood at 47: women

less intelligent women. In contrast, more intelligent men, despite having wanted to have fewer children at age 23, do not actually have fewer children by age 47. Among women, childhood general intelligence significantly decreases the number of children they have had in their lifetimes. Among men, it does not. While the effect of childhood general intelligence on women's fertility is consistent with the prediction of the Intelligence Paradox, the lack of the same effect among men is inconsistent with it.

As you can see in Figures 12.3 and 12.4, more intelligent women are significantly more likely to remain childless—and significantly less likely to become parents—than less intelligent women. The mean childhood IQ of women who have remained childless for life is 105.3, whereas the mean childhood IQ of women who have become parents is 101.7. The difference in mean childhood IQ between the two categories of women is very large and statistically significant.

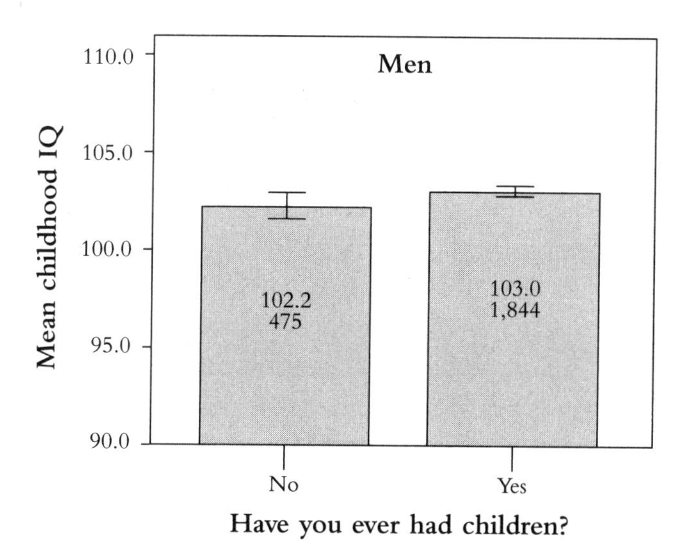

Figure 12.4 Association between mean childhood IQ and parenthood at 47: men

In contrast, more intelligent men are no more likely to remain childless for life than less intelligent men. The mean childhood IQ of men who have remained childless is 102.2 while the mean childhood IQ of men who have become parents is 103.0. The difference is not statistically significant. Men who remained childless and men who have become parents have essentially the same mean childhood IQ. This is once again contrary to the prediction of the Intelligence Paradox.

Why Women, Not Men?

It is not clear to me why more intelligent men, who wanted fewer children than less intelligent men at the start of their reproductive careers, do not actually have fewer children. This is in sharp contrast to more intelligent women who wanted fewer children and in fact do have fewer children than less intelligent women. The data, however, allow me to rule out some possible explanations.

Some might suspect that more intelligent women have fewer children than less intelligent women, because more intelligent women are more likely to pursue higher education or more demanding careers, and, in doing so, must sacrifice and forgo motherhood. In this view, women often have to make a difficult decision between family and careers, as the latter often demands pursuit of graduate education and investment in careers during the prime reproductive years for women. As a result, women cannot often pursue both family and careers and must choose between them. In contrast, men do not have to make such a decision. They can still pursue higher education or demanding careers, and at the same time manage to have family and children.

However, the data analysis shows that this is decidedly *not* the reason we observe the sex difference in the effect of childhood general intelligence on lifetime fertility among the NCDS respondents. It is not why more intelligent women have fewer children while more intelligent men don't. How do we know?

We know this because neither education nor income has any effect on the number of children women have. If the alternative explanation above is correct, then more educated women and women with greater earnings (which usually accompany demanding careers) should have fewer children than women with less education and earnings. But this is not the case. *Only* childhood intelligence, *not* educational achievement or earnings, decreases the number of children women have. Contrary to popular belief, *more educated women and women with more demanding careers do not have fewer children and are not more likely to remain childless.*

Another possibility is that women find intelligent men more attractive as mates. The evolutionary psychologist Geoffrey F. Miller has consistently argued that women preferentially select men with higher levels of intelligence to mate with.[2] Given that, as I discuss in Chapter 7, mating among mammalian species is largely a female choice, women's preference for intelligent men as mates can potentially explain why more intelligent men may end up with just as many children as less intelligent men, despite their desire to have fewer children and to remain childless.

There appears to be some evidence for this suggestion. Net of the same control variables as above, more intelligent men are significantly more likely to have ever been married and to be currently married at age 47 than less intelligent men. One standard deviation (15 IQ points) increase in childhood IQ increases men's odds of having ever been married by 23% and the odds of being currently married by 27%. In contrast, more intelligent women are no more likely to have ever been married, although they are more likely to be currently married. It is important to point out, however, that whether they have ever been married and whether they are currently married are always controlled in all the multiple regression analyses summarized above. So more intelligent men's greater tendency to have ever been married and to be currently married is not the only reason for the absence of the effect of intelligence on men's fertility and parenthood.

At any rate, the divergent effect of childhood intelligence on completed fertility for men and women, where more intelligent women have fewer children than less intelligent women but more intelligent men do not have fewer children than less intelligent men, means, among other things, that modern British people are not very endogamous on intelligence. More intelligent men do not appear to marry more intelligent women in the contemporary United Kingdom. If they are strongly endogamous on intelligence, then the fact that more intelligent women have fewer children will necessarily mean that more intelligent men also have fewer children. This is not the case, and the only reason is that more intelligent men are *not* married to more intelligent women, and vice versa.[3]

Heritability of Fertility: An Evolutionary Puzzle

Among both men and women, number of siblings significantly increases the number of children. Since the number of siblings (plus one) is the same as the number of children that *their parents* had, this means that fertility—the total number of children individuals have—is highly heritable. The more children your parents have had (and hence the more siblings you have), the more children you have yourself.

While this is consistent with earlier studies,[4] heritability of fertility—where the number of children the parents had is positively associated with the number of children the children have—is a mystery from an evolutionary psychological perspective.[5] The evolutionary psychological logic suggests that the association should be negative, not positive.

As I mention earlier in this chapter, there are two principal means for you to increase your reproductive success: you can have and raise your own children, or you can invest in close genetic relatives, such as your full siblings, who share half of your genes. But the option of maximizing reproductive success by investing in siblings is available only to those who have a large number of siblings. If you do not have any siblings, this option is not available to you. So you must have more children yourself if you do not have many siblings.

In contrast, if you do have a large number of siblings, then you don't necessarily have to have many children yourself, because you can increase your reproductive success by investing in your siblings. It is important to remember that *you share just as many genes with your full siblings as you do with your own biological children*. Both share half of your genes. Genetically speaking, it is slightly more advantageous to invest in your own children than in your full siblings, because your children belong in the next generation, not in the current generation with you and your siblings, and your children are expected to outlive your younger siblings.

Nevertheless, it is possible to increase your reproductive success by investing in your siblings.

In Chapter 3, I discuss a fundamental principle of quantitative genetics: that there is an inverse relationship between the heritability of a trait and its adaptiveness; the more adaptive a trait, the less heritable it is.[6] Fertility—the desire for and tendency to have children—is very adaptive; in fact, it's the very definition of reproductive success. So fertility should not be heritable at all, and all humans should be designed to have the maximum number of children that they can safely raise to sexual maturity, regardless of how many siblings they have.

If the number of siblings does affect fertility at all, then, for reasons I state above, evolutionary logic suggests that people who have more siblings should have fewer children (and instead invest in their siblings), and people who have fewer siblings should have more children. So the number of siblings and the number of children should be *negatively* correlated. Yet all the evidence suggests that this is not the case. Fertility appears to be heritable, and people who have more siblings have more children themselves. This remains an evolutionary puzzle.

Possible Societal Consequences

The analysis of the NCDS data suggests that more intelligent women are more likely to remain childless for life and to have fewer children than less intelligent women. If this finding is robust, and is true not only in the United Kingdom but in other western societies as well, what would it mean for these societies? What are the likely consequences of more intelligent women being more likely to remain childless and having fewer children than less intelligent women?

As I explain in Chapter 3, general intelligence is known to be highly heritable.[7] Genes determine about 80% of the variance in adult intelligence. On average, more intelligent parents beget more intelligent children. And the genes that influence general intelligence are thought to be located on the X chromosomes.[8] (In fact, as I mention in Chapter 9, they are thought to be located in the same region of the X chromosomes as the genes for male homosexuality, Xq28, which may or may not explain why homosexuals are more intelligent.) It means that boys inherit their general intelligence from their mothers only, while girls inherit their general intelligence from both their mothers and their fathers. Their fathers in turn inherit *their* general intelligence from *their* mothers (the girls' paternal grandmothers) only.

So women influence the general intelligence of future generations very strongly, through their sons and through their paternal granddaughters. If more intelligent women have fewer children and are more likely to remain childless, then one potential consequence is that the average level of general intelligence in society may decline over time.

Throughout the 20th century, the average level of general intelligence in most western industrial nations steadily increased. This phenomenon is now widely known as "the Flynn Effect,"[9] after two comprehensive reviews of secular increases in average IQ in many western industrialized nations conducted by James R. Flynn.[10] However, Richard Lynn documented the secular rise in intelligence in Japan a few years before Flynn did.[11] For this reason, some scientists prefer the name "the Lynn-Flynn Effect"[12] and I adopt the practice here. The first documentation of the secular rise in IQ may date even further back to the 1930s.[13]

Although there is no consensus on what caused the Lynn-Flynn Effect, one likely factor, identified by Richard Lynn himself,[14] is the increasing levels of infant and child nutrition and health. Regardless of their genetic endowment, healthier and better nourished babies on average grow up to have higher intelligence later in life than ill and malnourished babies. These factors likely more than compensated for the dysgenic fertility—where less intelligent parents have more children—throughout the 20th century, and the average level of intelligence has increased in most advanced industrial nations with the improved level of infant health and nutrition.

The improved health and nutrition as potential causes of the Lynn-Flynn Effect, however, would predict that the secular increase in general intelligence would halt in advanced industrial nations. The optimal level of health and nutrition has long passed and now obesity and diabetes have become serious problems in such nations. We are no longer getting healthier and better nourished; we are simply getting fatter in the United States and in other industrial nations.

If improved health and nutrition are chiefly responsible for the secular increase in general intelligence throughout the 20th century, and if these factors no longer contribute to the increase, then the negative effect of the dysgenic fertility should lead to a *declining* level of average intelligence in advanced industrial nations.[15] This is in fact happening already. There is strong evidence to suggest that the Lynn-Flynn Effect was only a 20th-century phenomenon. It appears to have ended at the end of the 20th century in the most advanced industrial nations (with simultaneously the highest rates of obesity and diabetes). Studies suggest that the average level of intelligence has begun to decline at the beginning of the 21st century in such advanced industrial nations as Australia,[16] Denmark,[17] Norway,[18] and the United Kingdom.[19]

Chapter 13

Other Possible Consequences of Intelligence

In previous chapters, I explain why more intelligent individuals are more likely to be liberals and atheists, why more intelligent men (but not women) are more likely to value sexual exclusivity (even though they may actually be *more* likely to have extramarital affairs), why night owls are more intelligent than morning larks, why homosexuals are more intelligent than heterosexuals, why more intelligent individuals prefer to listen to purely instrumental music such as classical music, why more intelligent individuals drink alcohol, smoke cigarettes, and use illegal drugs more, and why more intelligent women (but not men) have fewer children and are more likely to remain childless. All of these preferences, values, and lifestyles have one thing in common: they are all evolutionarily novel.

So what other preferences and values are evolutionarily novel? What else do more intelligent people like? What else can the Intelligence Paradox potentially explain?

Coffee

In Chapter 11, I discuss the effect of general intelligence on the consumption of alcohol, tobacco and illegal drugs. More intelligent people are more likely to consume these substances because they are evolutionarily novel.

The human consumption of coffee is even more recent in origin than that of alcohol or tobacco.[1] It is traced to Ethiopia in the 9th century.[2] The Intelligence Paradox would therefore predict that more intelligent individuals will consume more coffee than less intelligent individuals.

Among Add Health respondents in Wave I, those who usually have coffee or tea for breakfast on weekday mornings have a significantly (albeit very slightly) higher intelligence than those who don't (99.5 vs. 98.5). Net of age, sex, race, and religion, however, the effect of childhood intelligence on the consumption of coffee or tea is no longer statistically significant.

Vegetarianism

Another evolutionarily novel value is vegetarianism. Humans are naturally omnivorous, and anyone who eschewed animal protein and ate only vegetables in the ancestral environment, in the face of food scarcity and precariousness of its supply, was not likely to have survived long enough and stayed healthy enough to have left many offspring. So such a person is not likely to have become our ancestor. Anyone who preferentially ate animal protein and fat in the ancestral environment would have been much more

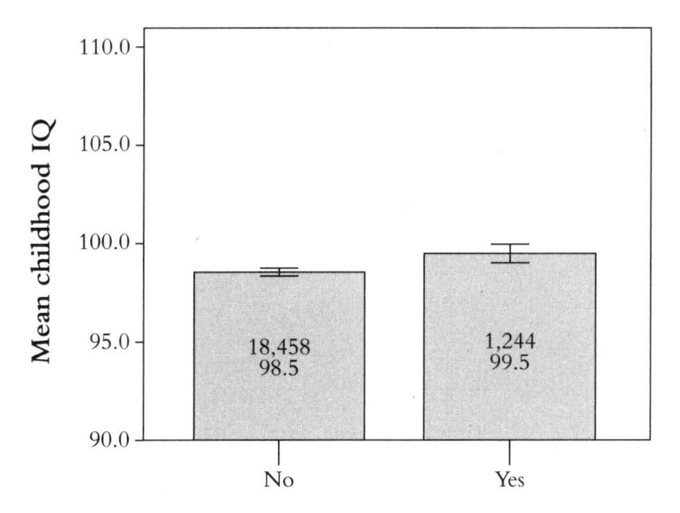

Figure 13.1 Association between childhood intelligence and consumption of caffeine (Add Health)

likely to live longer and stay healthier. They are therefore much more likely to have become our ancestors.

Vegetarianism would therefore be an evolutionarily novel value, as well as a luxury of abundance. The Intelligence Paradox would predict that more intelligent individuals are more likely to choose to become a vegetarian than less intelligent individuals.

This indeed appears to be the case.[3] Among the NCDS sample, those who are vegetarian at age 42 have significantly higher childhood general intelligence than those who are not vegetarian. Vegetarians have the mean childhood IQ of 109.1 whereas meat eaters have the mean childhood IQ of 100.9. The difference is large and highly statistically significant.

The association between childhood general intelligence and adult vegetarianism holds among both women and men separately. Among women, vegetarians have a mean childhood IQ of 108.0 while meat eaters have the mean childhood IQ of

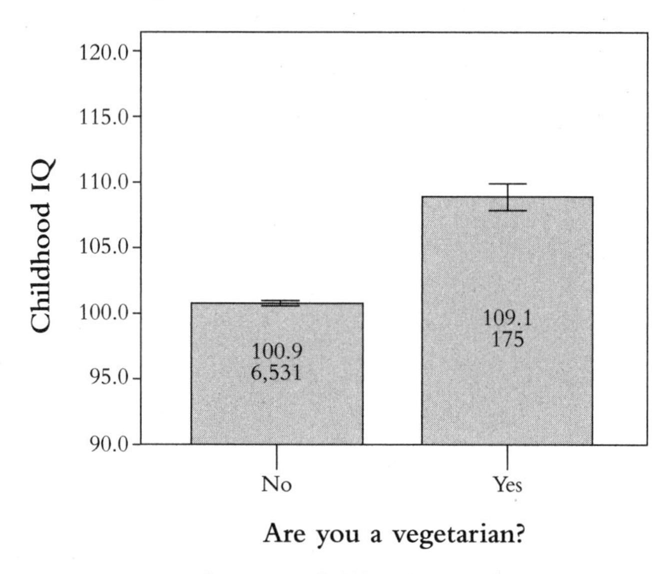

Figure 13.2 Association between childhood general intelligence and adult vegetarianism (NCDS)

100.7. Among men, vegetarians have the mean childhood IQ of 111.0 and meat eaters have the mean childhood IQ of 101.1, *a 10-point difference*!

The fact that the difference in childhood IQ between vegetarians and meat eaters is larger among men than among women makes sense in light of the historical division of labor between the sexes. Throughout evolutionary history, men have traditionally hunted animals for their meat while women have traditionally gathered plant food. So vegetarianism—a complete and total eschewal of animal meat—should be even more evolutionarily novel for men than for women. Women are 60% more likely to be vegetarians than men are (3.33% vs. 2.07%).

Childhood general intelligence has a significantly positive effect on the likelihood of vegetarianism at age 42, even net of a large number of social and demographic factors, such as sex, whether ever married, whether currently married, education,

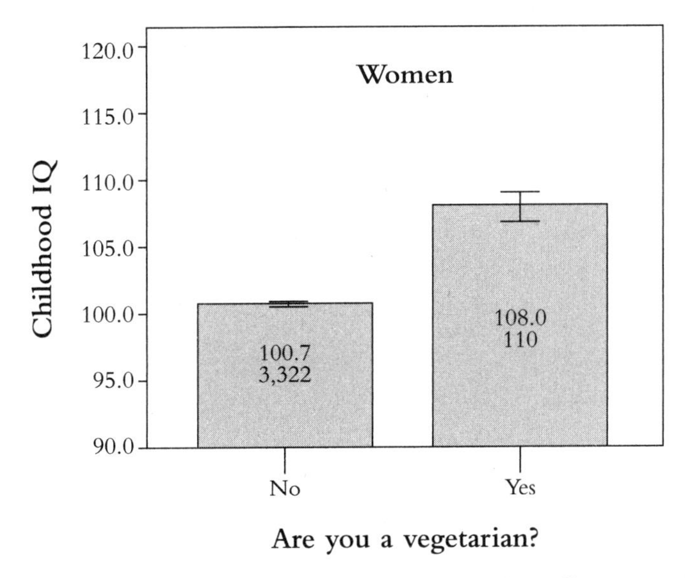

Figure 13.3 Association between childhood general intelligence and adult vegetarianism: women (NCDS)

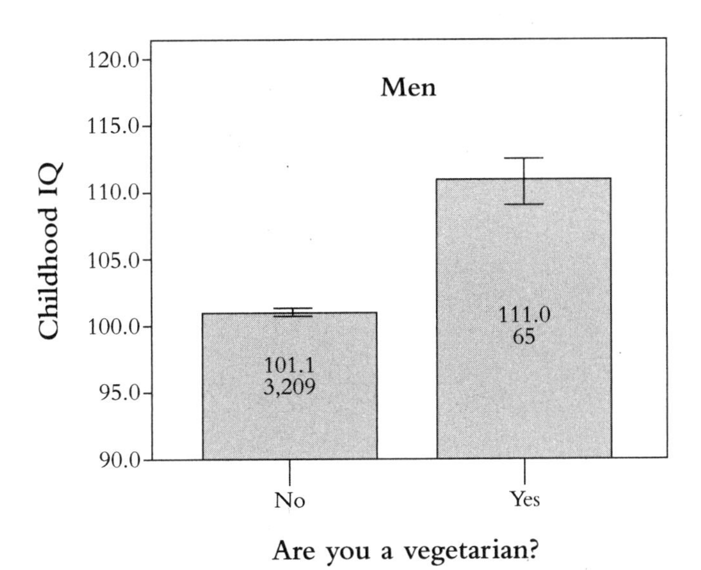

Figure 13.4 Association between childhood general intelligence and adult vegetarianism: men (NCDS)

income, religion, religiosity, social class at birth, mother's education, and father's education, both in the full sample and among men and women separately.

There appears very little doubt that more intelligent children are more likely to grow up to be vegetarian as adults in the United Kingdom. One standard deviation (15 points) increase in childhood IQ increases the odds of adult vegetarianism by 37% among women and by 48% among men.

Interestingly, the strong association between childhood intelligence and adult vegetarianism is not replicated with Add Health data in the US. American vegetarians in early adulthood *do* have significantly higher childhood intelligence in junior high and high school, but the difference is not large (101.5 vs. 99.3). And it is only significant among women (101.4 vs. 98.5), not among men (101.7 vs. 100.1). This is very strange given the historical division of labor by sex that I note above. The significant effect of

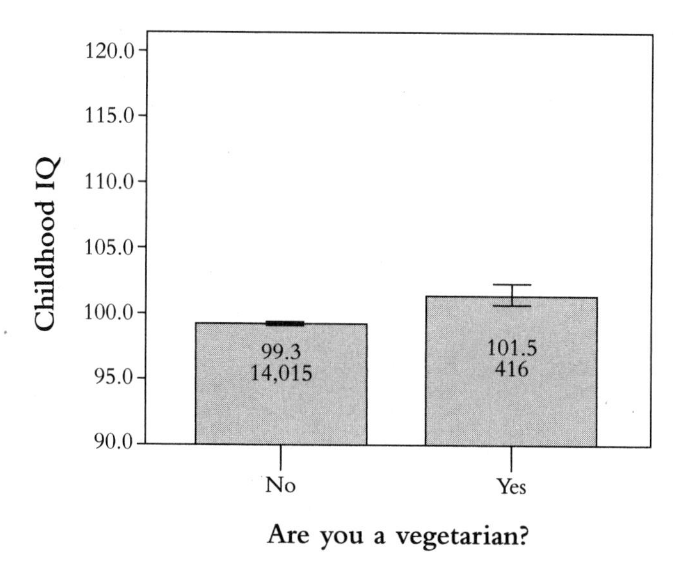

Figure 13.5 Association between childhood intelligence and adult vegetarianism (Add Health)

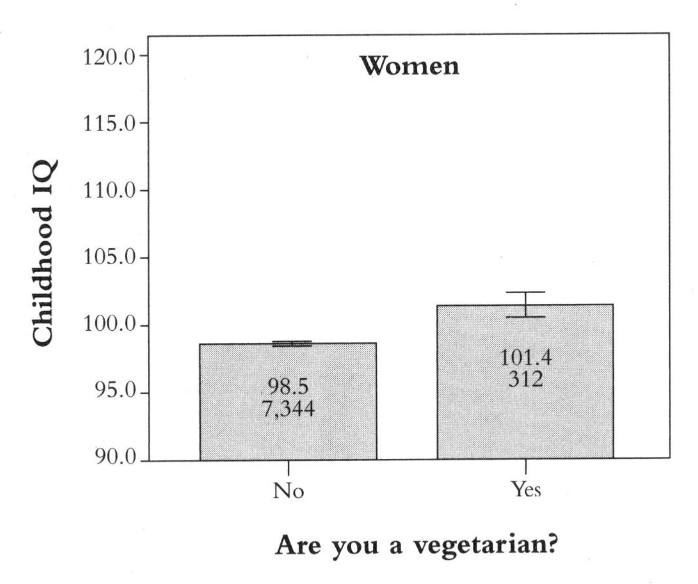

Figure 13.6 Association between childhood intelligence and adult vegetarianism: women (Add Health)

childhood intelligence on adult vegetarianism among Americans disappears entirely once mother's or father's education or religion is statistically controlled.

It is not at all clear to me why the difference in childhood intelligence between vegetarians and meat eaters is so much larger and stronger in the UK than in the US. As I note in Chapter 11, when I discuss the divergent effects of childhood intelligence on the adult consumption of tobacco, there are two principal differences between NCDS and Add Health: the national differences between the UK and the US; and the generational differences between those born in March 1958 and those born between 1974 and 1983. I am not sure if it is the national differences or the generational differences, or something entirely different, that account for the observed differences in the association between childhood intelligence and adult vegetarianism in the UK and the US.

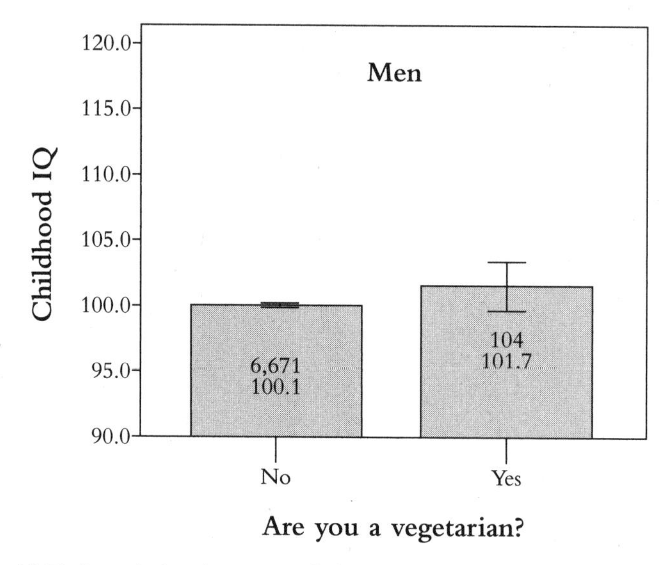

Figure 13.7 Association between childhood intelligence and adult vegetarianism: men (Add Health)

Crime and Punishment

As I note above in Chapter 11, the fact that criminals on average are less intelligent than noncriminals is consistent with the Intelligence Paradox. Much of what we call interpersonal crime today was probably a routine means of intrasexual male competition in the ancestral environment, so, in this sense, crime is "natural." In contrast, the technologies and institutions of law enforcement and criminal punishment are evolutionarily novel, so, in this sense, the police and the courts are "unnatural." It therefore makes sense from the perspective of the Intelligence Paradox that less intelligent men are more likely to resort to the "natural" means of criminal behavior to achieve their goals but they do not fully comprehend the "unnatural" entities of the criminal justice system.

Further, as I also explain in Chapter 11, what matters is not the criminality of the behavior per se, but its evolutionary novelty.

Less intelligent individuals are more likely to engage in evolutionarily familiar behavior and less likely to engage in evolutionarily novel behavior. This is why less intelligent individuals are *less* likely to consume illegal drugs because such consumption, while criminal, is evolutionarily novel.

The Intelligence Paradox would also suggest a novel hypothesis with regard to intelligence and criminality. As I mention in Chapter 11, while third-party enforcement (the police and the criminal justice system) is evolutionarily novel, second-party enforcement (retaliation and vigilance by the victims and their kin and allies) is not. So the Intelligence Paradox would predict that the difference in intelligence between criminals and noncriminals will disappear in situations where third-party enforcement of norms is weak or absent, and criminal behavior is largely controlled via second-party enforcement, such as situations of prolonged anarchy and statelessness, in fact, any situation that resembles the ancestral environment. If I'm right about this, it ironically means that less intelligent men will commit fewer crimes if the police disappeared, although more intelligent men may then commit more crimes. This may also explain why white-collar corporate crimes (of the types perpetrated by Enron and WorldCom), which are overwhelmingly committed by men of higher intelligence, abounded in the absence of the "police" under financial deregulation.

Representative Democracy

When Congressman Jack Murtha (D-PA) unexpectedly died in office in February 2010, one of the names brought up and briefly considered as a possible successor was his wife Joyce. This is very common. When Senator Ted Kennedy (D-MA) died in 2009, his wife Vicki was seriously considered as a possible interim successor. And when Congressman Sonny Bono (R-CA) died in office in 1998, his widow Mary did succeed him in office. She still represents her late husband's old district to this day, though now married to another Congressman from Florida (Connie Mack IV). All of this happened despite the fact that none of the widows had any political experience at the time of their husbands' deaths.

Sometimes political power is passed on to other family members. When my former Congressman Bud Shuster (R-PA) resigned in 2001 after a lengthy ethics investigation and a mild Congressional sanction,[4] he was succeeded in office by his son Bill, who still represents the 9th District of Pennsylvania to this day. His father's ethics problems apparently have not hampered his own success in politics.

And it turns out that the transmission of a Congressional seat from father to son has a long history in the United States. A *Time* magazine article from 1929 entitled "The Congress: Fathers & Sons," begins with the following (somewhat disturbing) paragraph.

> Primogeniture and hereditary public office have no place in U.S. tradition. This fact, however, did not last week deter the voters of the 7th Minnesota District from electing by a two-to-one majority Paul John Kvale (pronounced "Ka-volley") of Benson to the Congressional seat for six years occupied by his father, the Rev. Ole John Kvale, whose charred body was last month found in his burned summer cottage.[5]

Sometimes the son does not wait for the father to pass the torch to him. According to the Senate's own records, there has only been one father and son pair in history who served in the Senate simultaneously (Henry Dodge of Wisconsin from 1848 to 1857 and his son, Augustus Dodge of Iowa from 1855 to 1858).[6] Interestingly, there have been no father and son who served in the House of Representatives simultaneously, but there has been a mother and son pair (Frances Bolton and her son, Oliver Bolton, both of Ohio) who served simultaneously (1953–1957 and 1963–1965).[7]

And it is not limited to the U.S. Congress. We have elected family members as Presidents of the United States (the Adamses, the Harrisons, the Roosevelts, the Bushes, and almost the Kennedys and the Clintons). In Argentina, the popular President Néstor Kirchner chose not to seek (nearly certain) re-election for a second consecutive term and stepped aside in 2007, so that his wife Cristina could run and win the Presidency. And she did. Argentineans voted overwhelmingly for her and she won by a wide margin.

The United States is one of the oldest and most well established representative democracies in the world. It is also probably the only major world power which has never had any history of hereditary monarchy. In fact, the nation was founded with the very goal of rejecting the rule of hereditary monarchy. Why then, now that we have firmly established a secure form of representative democracy in the last two centuries, do we act as if we want hereditary monarchy, by electing wives, sons, and other family members of politicians to succeed?

Now, I'm sure that, just like any other profession or career, being a good politician requires certain skills and personality traits, and these skills and personality traits may very well be heritable. (Remember, Turkheimer's first law of behavior genetics?[8] All human traits are heritable. And the 50–0–50 rule, which I discuss in Chapter 3, suggests that many of these important traits may be 50% heritable.) So it makes sense that sons and other genetic relatives (*but not wives*) of former politicians want to pursue political careers and turn out to be good politicians themselves. Wives of politicians may also turn out to be good politicians themselves if there is assortative mating—where like marries like—on the important personality traits for politicians.

But that's *not* what I'm talking about. My question is, why do *the people* want the wives, sons, and other relatives of former politicians to succeed in office *and vote for them*, as if we have hereditary monarchy and politics ought to be family business?

Family business is ubiquitous. Everywhere in the world, sons and daughters inherit and continue their parents' occupations and professions. But politics in representative democracy is different because the continuation of family business requires popular support and consent. The son of the hardware store owner or the plastic surgeon does not require anyone's consent and support to continue his family business. The son of the Congressman does.

There exist family political dynasties in other democracies as well. For example, from 2005 to 2007, Lech Kaczynski was President of Poland while his twin brother Jaroslaw was Prime Minister.[9] If it turns out that people everywhere tend to want family members to succeed in political office, then such desire may very well be part of universal human nature. Does that mean that humans everywhere naturally want hereditary monarchy (but with popular support)? Is there something in our human nature that would want our political leaders to be succeeded by their wives, sons, and other family members?

People sometimes complain that the wives and the sons who inherit their political offices from their family members are not qualified to be elected. Such complaints were particularly strong for George W. Bush and Mary Bono. *But this is precisely the point.*

When a king dies, nobody asks the question "But is the crown prince *ready* and *qualified* to succeed to the throne?" Instead he automatically, unquestioningly, and immediately succeeds to his father's throne and becomes the next king, regardless of whether he is qualified or ready. Nobody complains that the legitimate son of a king is not qualified to succeed to the throne, because his bloodline *is* his qualification. That's how hereditary monarchy works.

My point is that we are acting like we are electing hereditary monarchs. Despite all the complaints about their utter lack of qualification, George W. Bush was reelected for the second term (a feat his father did not achieve), and Mary Bono continues to be reelected today. The fact that they and others may not be qualified for their office therefore supports my speculation.

If the desire for hereditary monarchy—political succession within the family—is part of human nature and universal among all humans, then it means that such a desire is evolutionarily familiar, and the desire for representative democracy—or any other form of government—is evolutionarily novel.

Our ancestors during most of human evolutionary history were undoubtedly more egalitarian and democratic than we were in the recent historical past, during the late agrarian and early industrial periods.[10] However, all the accoutrements of modern representative democracy—such as the secret ballot, one person-one vote, universal suffrage, and proportional representation—are all evolutionarily novel. The Intelligence Paradox would therefore suggest that more intelligent individuals and populations have greater desire and capacity for representative democracy than less intelligent individuals and populations.

Indeed this appears to be the case. In his comprehensive study of 170 nations in the world, the Finnish political scientist Tatu Vanhanen showed that the average intelligence in society increases its degrees of democracy.[11] The more intelligent the population on average, the more democratic their government. Vanhanen's finding suggests that representative democracy may indeed be evolutionarily novel and unnatural for humans. It does not necessarily mean that humans naturally prefer authoritarian government, the only major alternative form of government in the world today to representative democracy. After all, authoritarian government is also evolutionarily novel. My suggestion is merely that it may be natural for the human mind to expect their new political leader to be a blood relative of the old political leader, and that pure representative democracy, where political successors are not related to their predecessors, may therefore be unnatural.

However, recall from Chapter 1 the dangers of naturalistic fallacy. Natural does not mean good or desirable, and unnatural does not mean bad or undesirable. It simply means that humans did not evolve to practice representative democracy.

Conclusion

Intelligent People Are Not What You Think

By now I hope you have a very different view of intelligence and intelligent people than before you started reading this book. Yes, intelligent people have more education and do better in school, because formal education and universities, as well as many subjects taught in schools and universities, are entirely evolutionarily novel, although they may not be equally evolutionarily novel. Psychology (the study of human character and behavior) and home economics (the management of a household) are probably less evolutionarily novel than, say, trigonometry or particle physics. But studying any academic subject in school by listening to lectures, reading books, and taking written exams is evolutionarily novel.

In fact, probably all subjects taught in school are more or less evolutionarily novel, which is why we need to teach the students how to do them. We don't need to teach the students how to make friends, because it is part of human nature and everybody knows how to make friends on their own. Everybody, that is, except for intelligent people.

Yes, intelligent people make more money and attain higher status in organizations, because capitalist economy and complex organizations in which most people work today are entirely evolutionarily novel. Yes, intelligent people make better physicians, better astronauts, better scientists, and better violinists, because all of these pursuits are evolutionarily novel.

But these are all the unimportant things in life. We are not designed to be physicians, astronauts, scientists, or violinists. And intelligent people fail (or at least are no more successful than less intelligent people) in the most important things in life. They do not make better friends, they do not make better spouses and partners, and they do not make better parents, precisely because these are things that our ancestors have done for hundreds of thousands of years on the African savanna. Intelligent people—especially intelligent women—make the worst kind of parents, simply because they are least likely to *be* parents. And intelligent people lack common sense and have stupid ideas.

Think about it. If you had a choice, would you rather be a good brain surgeon, or a good parent? Would you rather be a good corporate executive, or a good friend?

I hope it is apparent by now that intelligence is just one of many, many traits that humans possess and on which there are individual differences, like height, weight, hair color, eye color, and many personality traits like aggressiveness and sociability. Just as intelligent people are different from less intelligent people—in both good and bad ways in most people's minds—taller people are different from shorter people, and sociable people are different from unsociable people. But we never equate any of these individual traits with human worth. We never believe that taller people or sociable people are inherently more worthy or better human beings than others who don't share their traits. (Yes, taller and better looking people make more money, but this is at least in part because they are on average more intelligent.[1]) That is why we don't get upset when there are observable group differences in these traits. Nobody gets upset that men are on average taller than women, and Caucasians are on average taller than Asians.

Yet, for some unfathomable reason, people treat intelligence differently. They believe (or at least act as though they believe) that intelligence is the ultimate gauge of human worth. They believe, or at least act publicly as though they believe, that everybody is or should be equally intelligent, because everybody is equally worthy as human beings. They get upset at scientific findings (by now, incontrovertible and indisputable) that show there are observable race and sex differences in intelligence. They *a priori* condemn such findings as racist or sexist. (Once again, as Kurzban aptly says, "It's only 'good science' if the message is politically correct."[2]) But they are no more racist than the finding that Caucasians are taller than Asians or blacks have higher blood pressure than whites; they are no more sexist than the finding that men are taller than women. Why should intelligence be any different?

Intelligent people are not at all what you think. Intelligent people are more likely to be liberal and atheistic. Why is it inherently better to be liberal and atheistic than to be conservative and religious? Intelligent people are more likely to be night owls than morning larks. Why is it inherently better to get up later in the morning (or in the afternoon) than earlier? Intelligent people are more likely to be homosexual. Why is it inherently better to be homosexual than heterosexual? Intelligent people prefer to listen to purely instrumental music than vocal music. Why is it inherently better to listen to classical music than folk music?

And more intelligent people are more likely to drink alcohol, to smoke tobacco, and to use illegal drugs. Intelligent people are more likely to binge drink and get drunk. Although this too is a value judgment, which science does not make, it seems very difficult to argue from any perspective that it is better to binge drink, get drunk, smoke cigarettes, and do drugs. Strictly from the health perspective, it is decidedly *not* good to drink alcohol (especially to excess), smoke tobacco, and use drugs.

And intelligent people—especially intelligent women—have fewer children and are more likely to remain childless for life than less intelligent people. Once again, whether or not to have children is a matter of personal choice, at least in western liberal societies, and it is neither better nor worse to have children than not to have children. Strictly from the perspective of your genes, however, not having children, or having fewer children than you can safely raise to sexual maturity, is the worst thing you can possibly do in your life. You are failing at the most important task in life, the one thing—the most important thing—that you are evolutionarily designed to do. More than exclusive homosexuality or listening to classical music, voluntary childlessness is the most unnatural thing that any living organism can do, including humans.

Reproductive success is the ultimate goal of all living organisms, including all humans. That is what all humans are evolutionarily designed to do. It is the meaning of life itself.[3] Voluntary childlessness is therefore the greatest crime against nature, which is why intelligent people do it.

Why is the tendency to commit the greatest crime against nature the ultimate gauge of human worth?

Notes

Introduction

1. Gardner (1983)..
2. Stigler and Becker (1977).
3. Friedman, Hechter and Kanazawa (1994).
4. Miller and Kanazawa (2007).
5. Herrnstein and Murray (1994); Murray (2003).
6. Davis, Smith, and Marsden (2009).
7. Huang and Hauser (1998); Miner (1957); Wolfle (1980).
8. http://www.cpc.unc.edu/projects/addhealth
9. Stanovich, Cunningham, and Feeman (1984); Zagar and Mead (1983).

1. What Is Evolutionary Psychology?

[1]. In this chapter, I will briefly introduce you to the field of evolutionary psychology. For a more comprehensive introduction, I (prejudicially) recommend that you turn to my earlier book *Why Beautiful People Have More Daughters* (Miller and Kanazawa, 2007), which is a comprehensive introduction to the entire field of evolutionary psychology written for a general, nonacademic audience. It's not only about why beautiful people have more daughters. You may also consult any of the other popular general introductions to the field, such as Matt Ridley's *The Red Queen* (1993), Robert Wright's *The Moral Animal* (1994), David M. Buss's *The Evolution of Desire* (1994), and Steven Pinker's *The Blank Slate* (2002).

2. Dunbar (1992).
3. Wilson (1975).
4. Barkow, Cosmides and Tooby (1992).
5. Betzig (1997a).

[6]. Culture: McGrew (1998), Wrangham et al. (1994); language: Reiss and McCowan (1993); Savage-Rumbaugh and Lewin (1994); tool use: van Lawick-Goodall (1964, 1968); consciousness: Gallup (1970); Plotnik, de Waal and Reiss (2006); Reiss and Marino (2001); morality: Brosnan and de Waal (2003); sympathy and compassion: de Waal (1996); romantic love: Leighton (1987); Smuts (1985); homosexuality: de Waal (1995); Bagemihl (2000); murder: Goodall (1986); Wrangham and Peterson (1996); rape: Thornhill (1980); Thornhill and Palmer (2000); Wrangham and Peterson (1996, pp. 132–143).

7. van den Berghe (1990, p. 428).
8. Campbell (1999, p. 243).
9. de Waal (1996); Brosnan and de Waal (2003).
10. Pinker (2002).
11. Endorsement on the cover of Betzig (1997b).
12. Ridley (1999, pp. 256–258).
13. Ellis (1996).
14. Moore (1903).
15. Hume (1739).
16. Davis (1978).
17. Ridley (1996, pp. 256–258).
18. http://www.epjournal.net/blog/2010/11/its-only-good-science-if-the-message-is-politically-correct/
19. Buss (2005).
20. Thornhill and Palmer (2000); Thornhill and Thornhill (1983).
21. http://www.psychologytoday.com/blog/the-scientific-fundamentalist
22. Kanazawa (2006a). http://www.psychologytoday.com/blog/the-scientific-fundamentalist/200802/if-the-truth-offends-it-s-our-job-offend

2. The Nature and Limitations of the Human Brain

1. Kanazawa (2004a).
2. Tooby and Cosmides (1990).

[3]. Cosmides and Tooby (1999, pp. 17–19); Shepard (1994). Some vision researchers disagree, however, and claim that perception of color is distorted under moonlight (Khan and Pattanaik, 2004) or reduced illumination (Shin, Yaguchi and Shioiri, 2004). If so, this is potentially due to the fact that our ancestors did not engage in many nocturnal activities (see Chapter 8). They likely woke up when the sun rose,

and went to sleep when the sun set. If the human vision system has difficulty accurately perceiving color under moonlight or reduced illumination, it may be because the need to do so is evolutionarily novel.

4. Crawford (1993); Symons (1990); Tooby and Cosmides (1990).
5. Kanazawa (2004a).
6. Burnham and Johnson (2005, pp. 130–131).
7. Hagan and Hammerstein (2006, pp. 341–343).

[8]. Most evolutionary psychologists and biologists concur that humans have not undergone significant evolutionary changes in the last 10,000 years, since the end of the Pleistocene Epoch, as the environment during this period has not provided a stable background against which natural and sexual selection could operate over many generations (Miller and Kanazawa, 2007, pp. 25–28). Evolution cannot operate against a moving target. This is the assumption behind the Savanna Principle. More recently, however, some scientists have voiced opinions that human evolution has continued and even accelerated during the Holocene Epoch in the last 10,000 years (Cochran and Harpending, 2009; Evans et al., 2005). While their studies conclusively demonstrate that new alleles (varieties of genes) have indeed emerged in the human genome since the end of the Pleistocene Epoch, the implications and importance of such new alleles for evolutionary psychology are not immediately obvious. In particular, *with the sole exception of lactose tolerance*, it is not clear whether these new alleles have led to the emergence of new evolved physiological and psychological adaptations in the last 10,000 years.

9. Kanazawa (2002).
10. Derrick, Gabriel and Hugenberg (2009).
11. http://www.imdb.com/name/nm0000210/bio
12. Held (2006); Malamuth (1996); Symons (1979, pp. 170–184).
13. Pérusse (1993, pp. 273–274).
14. Kenrick et al. (1989).
15. Ellis and Symons (1990).
16. Clark and Hatfield (1989); Hald and Høgh-Olesen (2010).
17. Buss and Schmitt (1993).
18. Sally (1995).

[19]. Fehr and Henrich (2003) suggest that one-shot encounters and exchanges might have been common in the ancestral environment. In their response to Fehr and Henrich, Hagen and Hammerstein

(2006) point out that, even if *one-shot* encounters were common in the ancestral environment, *anonymous* encounters could not have been common, and the game-theoretic prediction of defection in one-shot games requires both noniteration and anonymity. A lack of anonymity can lead to reputational concerns even in nonrepeated exchanges.

The available molecular genetic evidence suggests that our ancestors practiced female exogamy. It means that, when girls reached puberty, they left their natal group to marry into neighboring groups, in order to avoid inbreeding, whereas boys stayed in their natal group their entire lives. So all men in a hunter-gatherer band were genetically related to each other, whereas women were not. But they mostly stayed in the group they married into, so they became friends and allies for the rest of their lives. Seielstad, Minch and Cavalli-Sforza (1998).

20. van Beest and Williams (2006).
21. Eisenberger, Lieberman and Williams (2003).
22. Kanazawa (2006d).

3. What Is Intelligence?

1. Gottfredson (1997a); Neisser et al. (1996).

[2]. Earlier attempts by intelligence researchers to dispel similar misconceptions about intelligence include Gottfredson (1997b, 2009) and Herrnstein and Murray (1994, pp. 1–24).

3. Burt et al. (1995); Profant and Dimsdale (1999).
4. Gottfredson (1997b); Jensen (1980).
5. Jensen (1980).
6. Jensen (1998, pp. 49–50).

[7]. Technically, heritability is *the proportion of the variance* in the trait across individuals that is influenced by genes. In order to have nonzero heritability, there has to be variance in the trait across individuals. So even though genes *completely* and *entirely* determine how many eyes you have, heritability of the trait "number of eyes" is 0, because there is no variance in the trait across individuals. All normally developing humans have the identical number of eyes.

8. Turkheimer (2000).

[9]. Political attitudes: Alford, Funk and Hibbing (2005); Eaves and Eysenck (1974); divorce: Jockin, McGue and Lykken (1996); McGue and Lykken (1992).

10. http://www.psychologytoday.com/blog/the-scientific-fundamentalist/200809/the-50-0-50-rule-why-parenting-has-virtually-no-effect-chi
11. Harris (1995, 1998); Rowe (1994).
12. Bouchard et al. (1990); Rowe (1994).
13. Deary et al. (2004).
14. Silventoinen et al. (2003).
15. Falconer (1960).
16. Cosmides (1989).
17. Daly and Wilson (1988, pp. 37–93; 1999).
18. Chomsky (1957).
19. Kanazawa (2004b).
20. Barash (1982, pp. 144–147).
21. Orians and Heerwagen (1992).
22. Miller and Kanazawa (2007, pp. 25–28).

[23]. If the problem was novel but recurring from then on (which thus ceases to be novel), then there would eventually be an evolved psychological mechanism specifically to deal with it. General intelligence would not be necessary to solve such problems. General intelligence is necessary only for *novel* and *nonrecurrent* adaptive problems.

24. Gottfredson (1997a); Herrnstein and Murray (1994); Jensen (1998).
25. Herrnstein and Murray (1994).
26. Kanazawa (Forthcoming); Kanazawa and Perina (2009).
27. Jensen (1998, p. 296, Table 9.1).

4. When Intelligence Matters (and When It Doesn't)

1. Kanazawa (2010a).
2. Kanazawa (2002).
3. Kanazawa (2006b).

[4]. Romero and Goetz (2010). In addition to their empirical support of the Hypothesis, Romero and Goetz invent a handy acronym for the "Savanna-IQ Interaction Hypothesis"—SIQXH, which I happen to like very much.

5. Lubinski et al. (2006).
6. Frey and Detterman (2004); Kanazawa (2006e).
7. http://www.actstudent.org/faq/answers/actsat.html (emphases added)
8. Herrnstein and Murray (1994).

9. Herrnstein and Murray (1994, p. 168–172).
10. Herrnstein and Murray (1994, pp. 213–218).
11. Herrnstein and Murray (1994, pp. 225–229).
12. Crow (2003).
13. Herrnstein and Murray (1994, p. 216).
14. Deary (2008); Gottfredson and Deary (2004); Kanazawa (2006c).
15. Volk and Atkinson (2008).
16. Kanazawa (2005).
17. de Waal (1982).
18. Hamilton (1964).
19. Kanazawa (2004, p. 518).
20. Silverman et al. (2000).
21. Weiss, Morales and Jacobs (2003).
22. Romero and Goetz (2010).
23. Hall et al. (2008); Kingma et al. (2009).
24. Kingma et al. (2009).
25. Hall et al. (2008).
26. Kanazawa (2010a).

5. Why Liberals Are More Intelligent than Conservatives

1. Murray (1998).
2. Kanazawa (2010a).
3. Hamilton (1964).
4. Trivers (1971).
5. Whitmeyer (1997).
6. Whitmeyer (1997).
7. Dunbar (1992).
8. Levinson (1991–1995).
9. Chagnon (1992); Cronk (2004); Hill and Hurtado (1996); Lee (1979); Whitten (1976).
10. Ridley (1996).

[11]. See Note 19, Chapter 2, above.

12. Kanazawa (2010a).
13. Lake and Breglio (1992); Shapiro and Mahajan (1986); Wirls (1986).
14. Kluegel and Smith (1989); Sundquist (1983).
15. Deary, Batty, and Gale (2008).
16. Charlton (2009).
17. Miller and Kanazawa (2007, pp. 38–40).

18. Gallup (1990).
19. Zahavi (1975); Zahavi and Zahavi (1997).
20. Barkow (2006).
21. Kanazawa (2009).

6. Why Atheists Are More Intelligent than the Religious

1. Brown (1991).
2. Atran (2002); Boyer (2001); Guthrie (1993); Haselton and Nettle (2006); Kirkpatrick (2005).
3. Guthrie (1993).
4. Atran (2002).
5. Nesse (2001).
6. Nesse (2001).
7. Levinson (1991–1995).
8. Chagnon (1992); Cronk (2004); Hill and Hurtado (1996); Lee (1979); Whitten (1976).
9. Kanazawa (2010a).
10. Miller and Hoffman (1995); Miller and Stark (2002).
11. Kanazawa (2009); Lynn, Harvey and Nyborg (2009).

7. Why More Intelligent Men (but Not More Intelligent Women) Value Sexual Exclusivity

1. Miller and Kanazawa (2007, p. 81).
2. Miller and Kanazawa (2007, pp. 81–85). http://www.psychologytoday.com/blog/the-scientific-fundamentalist/200806/why-are-there-virtually-no-polyandrous-society-0
3. Alexander et al. (1977); Leutenegger and Kelly (1977).
4. Alexander et al. (1977); Leutenegger and Kelly (1977).
5. Harvey and Bennet (1985); Kanazawa and Novak (2005); Pickford (1986).
6. Kanazawa and Novak (2005).
7. Daly and Wilson (1988, pp. 140–142).
8. Kanazawa and Novak (2005).
9. Alexander et al. (1977, pp. 424–425, Table 15-1).
10. Eveleth and Tanner (1976).
11. Smith (1984).
12. Kanazawa (2001).
13. Baker and Bellis (1995); Gallup et al. (2003).

14. http://www.metro.co.uk/lifestyle/815295-cheat-on-wives-men-less-intelligent
15. http://articles.nydailynews.com/2010-03-02/entertainment/27057710_1_iqs-monogamy-smart-men
16. http://www.telegraph.co.uk/science/science-news/7339654/Intelligent-men-less-likely-to-cheat.html
17. Trivers (1972).
18. Clark and Hatfield (1989).
19. Hald and Høgh-Olesen (2010).
20. Gangestad and Simpson (2000).
21. Kancshiro ct al. (2008).
22. Dixson et al. (2010); Furnham, Tan and McManus (1997); Henss (2000); Singh (1993, 1994); Singh and Luis (1995); Singh and Young (1995); Tovée and Cornelissen (2001).
23. Kanazawa (2004b).
24. Buss (1994).
25. Kanazawa (2011); Kanazawa and Kovar (2004).
26. Kanazawa (2011).
27. http://www.psychologytoday.com/blog/the-scientific-fundamentalist/201012/beautiful-people-really-are-more-intelligent
28. Gangestad and Simpson (2000).
29. Jensen and Sinha (1993); Kanazawa and Reyniers (2009).
30. Buss and Schmitt (1993); Cameron, Oskamp and Sparks (1978); Lynn and Shurgot (1984); Gillis and Avis (1980).
31. Kanazawa (2009).
32. Kanazawa and Still (1999).
33. Shaw (1957, p. 254).

8. Why Night Owls Are More Intelligent than Morning Larks

1. http://www.psychologytoday.com/blog/the-scientific-fundamentalist/200809/the-50-0-50-rule-in-action-partisan-attachment
2. Fowler and Dawes (2008).
3. Kanazawa (1998, 2000).
4. Turkheimer (2000).
5. Alford, Funk and Hibbing (2005); Eaves and Eysenck (1974).
6. Bouchard et al. (1999); Koenig et al (2005).
7. Kanazawa (2010a).
8. Vitaterna, Takahashi and Turek (2001, p. 85).
9. Klein, Moore and Reppert (1991).

10. King and Takahashi (2000).
11. Hur (2007).
12. Vitaterna et al. (2001, p. 90, emphasis added).
13. Dyer et al. (2009, p. 8964); Ross (2000).
14. Levinson (1991–1995).
15. Chagnon (1992); Cronk (2004); Hill and Hurtado (1996); Lee (1979); Whitten (1976).
16. Chagnon (1992, p. 129).
17. Cronk (2004, p. 88).
18. Chagnon (1992, p. 132).
19. Cronk (2004, p. 93).
20. Hill and Hurtado (1996, p. 65).
21. Cronk (2004, p. 92).
22. Ash and Gallup (2007); Bailey and Geary (2009); Kanazawa (2008).
23. Kanazawa and Perina (2009).
24. Roberts and Kyllonen (1999).
25. Kanazawa and Perina (2009).

9. Why Homosexuals Are More Intelligent than Heterosexuals

1. Bagemihl (2000); de Waal (1995).
2. Kirk et al. (2000).
3. Bailey and Pillard (1991).
4. Ellis and Ames (1987).
5. Blanchard and Bogaert (1996a); Bogaert (2003).
6. Mustanski, Chivers and Bailey (2002); Wilson and Rahman (2005).
7. Bailey (2009); Chivers et al. (2004); Diamond (2008).
8. Mustanski et al. (2002, pp. 122–127); Wilson and Rahman (2005, pp. 13–16).
9. Adams, Wright and Lohr (1996).
10. Wilson and Rahman (2005, p. 15).
11. Chivers, Seto and Blanchard (2007).
12. Hamer et al. (1993); Turner (1996a).
13. Levinson (1991–1995).
14. Levinson (1991–1995, Volume 2, p. 79).
15. Levinson (1991–1995, Volume 2, p. 285).
16. Chagnon (1992); Cronk (2004); Hill and Hurtado (1996); Lee (1979); Whitten (1976).
17. Hill and Hurtado (1996, pp. 276–277; emphasis added).
18. Hill and Hurtado (1996, p. 277).

19. Blanchard and Bogaert (1996b); Bogaert and Blanchard (1996).
20. Kanazawa (Forthcoming).
21. Kanazawa (Forthcoming).
22. Diamond (2008).
23. Kanazawa (Forthcoming).
24. Kanazawa (Forthcoming).

10. Why More Intelligent People Like Classical Music

1. http://76.74.24.142/8EF388DA-8FD3-7A4E-C208-CDF1ADE8B179.pdf
2. http://www.npr.org/about/aboutnpr/stations_publicmedia.html
3. http://www.npr.org/music/
4. Mithen (2005).
5. Brown (2000).
6. Bickerton (1990); Jackendoff (2000).
7. Wray (1998).
8. Mithen (2005, pp. 105–121).
9. Zuberbühler (2002, 2003).
10. Leinonen et al. (1991); Linnankoski et al. (1994).
11. Leinonen et al. (2003).
12. Mithen (2005, pp. 28–68).
13. Nettle (1983).
14. Everett (2005).
15. Everett (2005, p. 622).
16. Wray (2006, p. 104).
17. Fry (1948); Kalmus and Fry (1980).
18. Rentfrow and Gosling (2003).
19. Kanazawa and Perina (forthcoming).
20. Kanazawa (2006b).
21. Mithen (2005).
22. Rentfrow and Gosling (2003).

11. Why Intelligent People Drink and Smoke More

1. Dudley (2000).
2. Vallee (1998).
3. Vallee (1998, p. 81).
4. Dudley (2000, p. 9).
5. Dudley (2000).
6. Dudley (2000, p. 9).

7. Goodspeed (1954).
8. Wilbert (1991).
9. Goodman (1993); Smith (1999).
10. Smith (1999, p. 377).
11. Brownstein (1993).
12. Smith (1999, pp. 381–382).
13. Smith (1999).
14. Smith (1999).
15. Holmstedt and Fredga (1981).
16. Johnson et al. (2009); Batty et al. (2007); Batty, Deary and Macintyre (2007).
17. Herrnstein and Murray (1994); Hirschi and Hindelang (1977); Wilson and Herrnstein (1985).
18. Wolfgang, Figlio and Sellin (1972); Yeudall, Fromm-Auch and Davies (1982).
19. Gibson and West (1970).
20. Wolfgang et al. (1972); Moffitt (1990).
21. Lynam, Moffitt and Stouthamer-Loeber (1993); Moffitt et al. (1981).
22. Moffitt and Silva (1988).
23. Ellis (1998).
24. de Waal (1998).
25. de Waal (1992).
26. de Waal, Luttrel and Canfield (1993).

12. Why Intelligent People Are the Ultimate Losers in Life

1. Fielder and Huber (2007).
2. Miller 2000a, 2000b, 2009.

[3]. I owe this insight to Lena Edlund.

4. Kohler, Rodgers and Christensen (1999); Kohler et al. (2006).
5. Miller and Kanazawa (2007, p. 184).
6. Falconer (1960).
7. Jensen (1998).
8. Lehrke (1972, 1997); Turner (1996a, 1996b).
9. Herrnstein and Murray (1994, pp. 307–309).
10. Flynn (1984, 1987).
11. Lynn (1982).
12. Murphy (2011); te Nijenhuis (2011).
13. Merrill (1938).
14. Lynn (1990, 1998).

15. Lynn and Harvey (2008).
16. Cotton et al. (2005).
17. Teasdale and Owen (2008).
18. Sundet, Barlaug and Torjussen (2004).
19. Shayer, Ginsburg and Coe (2007); Shayer and Ginsburg (2009).

13. Other Possible Consequences of Intelligence

1. Smith (1999).
2. Pendergrast (1999).
3. Gale et al. (2007).
4. http://www.nytimes.com/2000/10/06/us/congressman-draws-rebuke-but-no-penalty.html
5. http://www.time.com/time/magazine/article/0,9171,752267,00.html. Yes, a *Time* article published in 1929 is available online! I have no idea why.
6. http://www.senate.gov/artandhistory/history/common/briefing/Facts_Figures.htm
7. http://clerk.house.gov/art_history/house_history/family_firsts.html
8. Turkheimer (2000).
9. Nizynska (2010); http://www.time.com/time/magazine/article/0,9171,1655447,00.html
10. Boehm (1999); Lenski (1966).
11. Vanhanen (2003).

Conclusion: Intelligent People Are Not What You Think

1. Jensen and Sinha (1993); Kanazawa (2011); Kanazawa and Kovar (2004); Kanazawa and Reyniers (2009).
2. http://www.epjournal.net/blog/2010/11/its-only-good-science-if-the-message-is-politically-correct/
3. Dennett (1995); Kanazawa (2004c).

References

Adams, Henry E., Lester W. Wright, and Bethany A. Lohr. 1996. "Is Homophobia Associated with Homosexual Arousal?" *Journal of Abnormal Psychology.* 105: 440–445.

Alexander, Richard D., John L. Hoogland, Richard D. Howard, Katharine M. Noonan, and Paul W. Sherman. 1979. "Sexual Dimorphisms and Breeding Systems in Pinnipeds, Ungulates, Primates and Humans." Pp. 402–435 in *Evolutionary Biology and Human Social Behavior: An Anthropological Perspective*, edited by Napoleon A. Chagnon and William Irons. North Scituate, Mass.: Duxbury Press.

Alford, John R., Carolyn L. Funk, and John R. Hibbing. 2005. "Are Political Orientations Genetically Transmitted?" *American Political Science Review.* 99: 153–167.

Ash, Jessica and Gallup, Gordon G. Jr., 2007. "Paleoclimatic Variation and Brain Expansion during Human Evolution." *Human Nature.* 18: 109–124.

Atran, Scott. 2002. *In Gods We Trust: The Evolutionary Landscape of Religion.* Oxford, U.K.: Oxford University Press.

Bagemihl, Bruce. 2000. *Biological Exuberance: Animal Homosexuality and Natural Diversity.* New York: St. Martin's Press.

Bailey, Drew H. and David C. Geary. 2009. "Hominid Brain Evolution: Testing Climatic, Ecological, and Social Competition Models." *Human Nature.* 20: 67–79.

Bailey, J. Michael. 2009. "What Is Sexual Orientation and Do Women Have One?" *Nebraska Symposium on Motivation.* 54: 43–63.

Bailey, J. Michael and Richard C. Pillard. 1991. "A Genetic Study of Male Sexual Orientation." *Archives of General Psychiatry.* 48: 1089–1096.

Baker, R. Robin and Mark A. Bellis. 1995. *Human Sperm Competition: Copulation, Masturbation and Infidelity*. London: Chapman and Hall.

Barash, David P. 1982. *Sociobiology and Behavior, Second Edition*. New York: Elsevier.

Barkow, Jerome H. 2006. "Introduction: Sometimes the Bus Does Wait." Pp. 3–59 in *Missing the Revolution: Darwinism for Social Scientists*, edited by Jerome H. Barkow. Oxford: Oxford University Press.

Barkow, Jerome H., Leda Cosmides and John Tooby. (Editors.) 1992. *The Adapted Mind: Evolutionary Psychology and the Generation of Culture*. New York: Oxford University Press.

Batty, G. David, Ian J. Deary, and Linda S. Gottfredson. 2007. "Premorbid (Early Life) IQ and Later Mortality Risk: Systematic Review." *Annals of Epidemiology*. 17: 278–288.

Batty, G. David, Ian J. Deary, Ingrid Schoon, and Catharine R. Gale. 2007. "Mental Ability across Childhood in Relation to Risk Factors for Premature Mortality in Adult Life: The 1970 British Cohort Study." *Journal of Epidemiology and Community Health*. 61: 997–1003.

Betzig, Laura. 1997a. "People Are Animals." Pp. 1–17 in *Human Nature: A Critical Reader*, edited by Laura Betzig. New York: Oxford University Press.

Betzig, Laura. (Editor.) 1997b. *Human Nature: A Critical Reader.* New York: Oxford University Press.

Bickerton, Derek. 1990. *Language and Species.* Chicago: University of Chicago Press.

Blanchard, Ray and Anthony F. Bogaert. 1996a. "Homosexuality in Men and Number of Older Brothers." *American Journal of Psychiatry.* 153: 27–31.

Blanchard, Ray and Anthony F. Bogaert. 1996b. "Biodemographic Comparisons of Homosexual and Heterosexual Men in the Kinsey Interview Data." *Archives of Sexual Behavior.* 25: 551–579.

Boehm, Christopher. 1999. *Hierarchy in the Forest: The Evolution of Egalitarian Behavior.* Cambridge: Harvard University Press.

Bogaert, Anthony F. 2003. "Number of Older Brothers and Sexual Orientation: New Tests and the Attraction/Behavior Distinction in Two National Probability Samples." *Journal of Personality and Social Psychology.* 84: 644–652.

Bogaert, Anthony F. and Ray Blanchard. 1996. "Physical Development and Sexual Orientation in Men: Height, Weight and Age of Puberty Differences." *Personality and Individual Differences.* 21: 77–84.

Bouchard, Thomas J. Jr., David T. Lykken, Matthew McGue, Nancy L. Segal, and Auke Tellegen. 1990. "Sources of Human Psychological Differences: The Minnesota Study of Twins Reared Apart." *Science.* 250: 223–228.

Bouchard, Thomas J. Jr., Matt McGue, David Lykken, and Auke Tellegen. 1999. "Intrinsic and Extrinsic Religiousness: Genetic and Environmental Influences and Personality Correlates." *Twin Research.* 2: 88–98.

Boyer, Pascal. 2001. *Religion Explained: The Evolutionary Origins of Religious Thought.* New York: Basic Books.

Brosnan, Sarah F. and Frans B. M. de Waal. 2003. "Monkeys Reject Unequal Pay." *Nature.* 425: 297–299.

Brown, Donald E. 1991. *Human Universals.* New York: McGraw-Hill.

Brown, Steven. 2000. "The 'Musilanguage' Model of Music Evolution." Pp. 271–300 in *The Origins of Music,* edited by Nils L. Wallin, Björn Merker, and Steven Brown. Cambridge: MIT Press.

Brownstein, Michael J. 1993. "A Brief History of Opiates, Opioid Peptides, and Opioid Receptors." *Proceedings of the National Academy of Sciences.* 90: 5391–5393.

Burnham, Terence C. and Dominic D. P. Johnson. 2005. "The Biological and Evolutionary Logic of Human Cooperation." *Analyse & Kritik.* 27: 113–135.

Burt, Vicki L., Paul Whelton, Edward J. Roccella, Clarice Brown, Jeffrey A. Cutler, Millicent Higgins, Michael J. Horan, and Darwin Labarthe. 1995. "Prevalence of Hypertension in the US Adult Population." *Hypertension.* 25: 305–313.

Buss, David M. 1994. *The Evolution of Desire: Strategies of Human Mating,* Second Edition. New York: Basic Books.

Buss, David M. 2005. *The Murderer Next Door: Why the Mind Is Designed to Kill.* New York: Penguin.

Buss, David M. and David P. Schmitt. 1993. "Sexual Strategies Theory: An Evolutionary Perspective on Human Mating." *Psychological Review.* 100: 204–232.

Cameron, Catherine, Stuart Oskamp, and William Sparks. 1977. "Courtship American Style: Newspaper Ads." *Family Coordinator.* 26: 27–30.

Campbell, Anne. 1999. "Staying Alive: Evolution, Culture, and Women's Intrasexual Aggression." *Behavior and Brain Sciences.* 22: 203–252.

Chagnon, Napoleon. 1992. *Yanomamö*, 4th ed. Fort Worth: Harcourt Brace Jovanovich.

Charlton, Bruce G. 2009. "Clever Sillies: Why High IQ People Tend to Be Deficient in Common Sense." *Medical Hypotheses.* 73: 867–870.

Chivers, Meredith L., Gerulf Rieger, Elizabeth Latty, and J. Michael Bailey. 2004. "A Sex Difference in the Specificity of Sexual Arousal." *Psychological Science.* 15: 736–744.

Chivers, Meredith L., Michael C. Seto, and Ray Blanchard. 2007. "Gender and Sexual Orientation Differences in Sexual Response to Sexual Activities versus Gender of Actors in Sexual Films." *Journal of Personality and Social Psychology.* 93: 1108–1121.

Chomsky, Noam. 1957. *Syntactic Structures.* The Hague: Mouton.

Clark, Russell D. III and Elaine Hatfield. 1989. "Gender Differences in Receptivity to Sexual Offers." *Journal of Psychology and Human Sexuality.* 2: 39–55.

Cochran, Gregory and Henry Harpending. 2009. *The 10,000 Year Explosion: How Civilization Accelerated Human Evolution.* New York: Basic Books.

Cosmides, Leda. 1989. "The Logic of Social Exchange: Has Natural Selection Shaped How Humans Reason? Studies with the Wason Selection Task." *Cognition.* 31: 187–276.

Cosmides, Leda and John Tooby. 1999. *What Is Evolutionary Psychology?* Center for Evolutionary Psychology. University of California at Santa Barbara.

Cotton, Sue M., Patricia M. Kiely, David P. Crewther, Brenda Thomson, Robin Laycock, and Sheila G. Crewther. 2005. "A Normative and Reliability Study for the Raven's Coloured Progressive Matrices for Primary School Aged Children from Victoria, Australia." *Personality and Individual Differences.* 39: 647–659.

Crawford, Charles B. 1993. "The Future of Sociobiology: Counting Babies or Proximate Mechanisms?" *Trends in Ecology and Evolution.* 8: 183–186.

Cronk, Lee. 2004. *From Mukogodo to Maasai: Ethnicity and Cultural Change in Kenya.* Boulder: Westview Press.

Daly, Martin and Margo Wilson. 1988. *Homicide.* New York: De Gruyter.

Daly, Martin and Margo Wilson. 1999. *The Truth about Cinderella: A Darwinian View of Parental Love.* New Haven: Yale University Press.

Davis, Bernard. 1978. "Moralistic Fallacy." *Nature.* 272: 390.

Davis, James Allan, Tom W. Smith, and Peter V. Marsden. 2009. *General Social Surveys, 1972–2008: Cumulative Codebook.* Chicago: National Opinion Research Center.

Deary, Ian J. 2008. "Why Do Intelligent People Live Longer?" *Nature.* 456: 175–176.

Deary, Ian J., G. David Batty, and Catharine R. Gale. 2008. "Childhood Intelligence Predicts Voter Turnout, Voting Preferences, and Political Involvement in Adulthood: The 1970 British Cohort Study." *Intelligence.* 36: 548–555.

Deary, Ian J., Martha C. Whiteman, John M. Starr, Lawrence J. Whalley, and Helen C. Fox. 2004. "The Impact of Childhood Intelligence on Later Life: Following Up the Scottish Mental Surveys of 1932 and 1947." *Journal of Personality and Social Psychology.* 86: 130–147.

Dennett, Daniel C. 1995. *Darwin's Dangerous Idea: Evolution and the Meanings of Life.* New York: Touchtone.

Derrick, Jaye L., Shira Gabriel, and Kurt Hugenberg. 2009. "Social Surrogacy: How Favored Television Programs Provide the Experience of Belonging." *Journal of Experimental Social Psychology.* 45: 352–362.

Diamond, Lisa M. 2008. *Sexual Fluidity: Understanding Women's Love and Desire.* Cambridge: Harvard University Press.

Dixson, Barnaby J., Alan F. Fixson, Phil J. Bishop, and Amy Parish. 2010. "Human Physique and Sexual Attractiveness in Men and Women: A New Zealand-U.S. Comparative Study." *Archives of Sexual Behavior.* 39: 798–806.

Dudley, Robert. 2000. "Evolutionary Origins of Human Alcoholism in Primate Frugivory." *Quarterly Review of Biology.* 75: 3–15.

Dunbar, Robin I. M. 1992. "Neocortex Size as a Constraint on Group Size in Primates." *Journal of Human Evolution.* 20: 469–493.

Dyer, Michael A., Rodrigo Martins, Manoel da Silva Filho, José Augusto P. C. Muniz, Luiz Carlos L. Silveira, Constance L. Cepko, and Barbara L. Finlay. 2009. "Developmental Sources of Conservation and Variation in the Evolution of the Primate Eye." *Proceedings of the National Academy of Science.* 106: 8963–8968.

Eaves, L. J. and H. J. Eysenck. 1974. "Genetics and the Development of Social Attitudes." *Nature.* 249: 288–289.

Eisenberger, Naomi I., Matthew D. Lieberman, and Kipling D. Williams. 2003. "Does Rejection Hurt?: An fMRI Study of Social Exclusion." *Science.* 302: 290–292.

Ellis, Lee. 1996. "A Discipline in Peril: Sociology's Future Hinges on Curing Its Biophobia." *American Sociologist*. 27: 21–41.

Ellis, Lee. 1998. "Neodarwinian Theories of Violent Criminality and Antisocial Behavior: Photographic Evidence from Nonhuman Animals and a Review of the Literature." *Aggression and Violent Behavior.* 3: 61–110.

Ellis, Lee and M. Ashley Ames. 1987. "Neurohormonal Functioning and Sexual Orientation: A Theory of Homosexuality-Heterosexuality." *Psychological Bulletin*. 101: 233–258.

Evans, Patrick D., Sandra L. Gilbert, Nitzan Mekel-Bobrov, Eric J. Wallender, Jeffrey R. Anderson, Leila M. Vaez-Azizi, Sarah A. Tishkoff, Richard R. Hudson, and Bruce T. Lahn. 2005. "*Microcephalin*, a Gene Regulating Brain Size, Continues to Evolve Adaptively in Humans." *Science*. 309: 1717–1720.

Eveleth, Phyllis B. and James M. Tanner. 1976. *Worldwide Variation in Human Growth*. Cambridge, U.K.: Cambridge University Press.

Falconer, Douglas S. 1960. *Introduction to Quantitative Genetics.* New York: Ronald Press.

Fehr, Ernst and Joseph Henrich. 2003. "Is Strong Reciprocity a Maladaptation? On the Evolutionary Foundations of Human Altruism." Pp. 55–82 in *Genetic and Cultural Evolution of Cooperation*, edited by Peter Hammerstein. Cambridge: MIT Press.

Fielder, Martin and Susanne Huber. 2007. "The Effects of Sex and Childlessness on the Association between Status and Reproductive Output in Modern Society." *Evolution and Human Behavior.* 28: 392–398.

Flynn, James R. 1984. "The Mean IQ of Americans: Massive Gains 1932 to 1978." *Psychological Bulletin*. 95: 29–51.

Flynn, James R. 1987. "Massive IQ Gains in 14 Nations: What IQ Tests Really Measure." *Psychological Bulletin*. 101: 171–191.

Fowler, James H. and Christopher T. Dawes. 2008. "Two Genes Predict Voter Turnout." *Journal of Politics.* 70: 579–594.

Frey, Meredith C. and Douglas K. Detterman. 2004. "Scholastic Assessment or *g*?: The Relationship Between the Scholastic Assessment Test and General Cognitive Ability." *Psychological Science*. 15: 373–378.

Friedman, Debra, Michael Hechter, and Satoshi Kanazawa. 1994. "A Theory of the Value of Children." *Demography.* 31: 375–401.

Fry, D. B. 1948. "An Experimental Study of Tone Deafness." *Speech*. 1948: 1–7.

Furnham, Adrian, Tina Tan, and Chris McManus. 1997. "Waist-to-Hip Ratio and Preferences for Body Shape: A Replication and Extension." *Personality and Individual Differences.* 22: 539–549.

Gale, Catharine R., Ian J. Deary, Ingrid Schoon, and G. David Batty. 2007. "IQ in Childhood and Vegetarianism in Adulthood: 1970 British Cohort Study." *British Medical Journal.* 334: 245–248.

Gallup, Gordon G. Jr. 1970. "Chimpanzees: Self-Recognition." *Science.* 167: 86–87.

Gallup, Gordon G. Jr. 1990. "Review of *The Unheeded Cry: Animal Consciousness, Animal Pain and Science*, by Gernard E. Rollin." *Animal Behaviour.* 40: 200–201.

Gallup, Gordon G. Jr., Rebecca L. Burch, Mary L. Zappieri, Rizwan A. Parvez, Malinda L. Stockwell, and Jennifer A. Davis. 2003. "The Human Penis as a Semen Displacement Device." *Evolution and Human Behavior.* 24: 277–289.

Gangestad, Steven W. and Jeffry A. Simpson. 2000. "The Evolution of Human Mating: Trade-offs and Strategic Pluralism." *Behavioral and Brain Sciences.* 23: 573–644.

Gardner, Howard. 1983. *Frames of Mind: The Theory of Multiple Intelligences.* New York: Basic Books.

Gibson, H. B. and D. J. West. 1970. "Social and Intellectual Handicaps as Precursors of Early Delinquency." *British Journal of Criminology.* 10: 21–32.

Gillis, John S. and Walter E. Avis. 1980. "The Male-Taller Norm in Mate Selection." *Personality and Social Psychology Bulletin.* 6: 396–401.

Goodall, Jane. 1986. *Chimpanzees of Gombe: Patterns of Behavior.* Cambridge: Harvard University Press.

Goodall, Jane. 2001. "Tool-Use and Aimed Throwing in a Community of Free-Living Chimpanzees." *Nature.* 201: 1264–1266.

Goodman, Jordan. 1993. *Tobacco in History: The Cultures of Dependence.* London: Routledge.

Goodspeed, Thomas Harper. 1954. *The Genus Nicotiana.* Waltham, Mass.: Chronica Botanica.

Gottfredson, Linda S. 1997a. "Why *g* Matters: The Complexity of Everyday Life." *Intelligence.* 24: 79–132.

Gottfredson, Linda S. 1997b. "Mainstream Science on Intelligence: An Editorial with 52 Signatories, History, and Bibliography." *Intelligence.* 24: 13–23.

Gottfredson, Linda S. 2009. "Logical Fallacies Used to Dismiss the Evidence on Intelligence Testing." Pp. 11–65 in *Correcting Fallacies about Educational and Psychological Testing*, edited by Richard P. Phelps. Washington, D.C.: American Psychological Association.

Gottfredson, Linda S. and Ian J. Deary. 2004. "Intelligence Predicts Health and Longevity, but Why?" *Current Directions in Psychological Science.* 13: 1–4.

Guthrie, Stewart Elliott. 1993. *Faces in the Clouds: A New Theory of Religion.* New York: Oxford University Press.

Hagen, Edward H. and Peter Hammerstein. 2006. "Game Theory and Human Evolution: A Critique of Some Recent Interpretations of Experimental Games." *Theoretical Population Biology.* 69: 339–348.

Hald, Gert Martin. 2006. "Gender Differences in Pornography Consumption among Young Heterosexual Danish Adults." *Archives of Sexual Behavior.* 35: 577–585.

Hald, Gert Martin and Henrik Høgh-Olesen. 2010. "Receptivity to Sexual Invitations from Strangers of the Opposite Gender." *Evolution and Human Behavior.* 31: 453–458.

Hall, Peter A., Lorin J. Elias, Geoffrey T. Fong, Amabilis H. Harrison, Ron Borowsky, and Gordon E. Sarty. 2008. "A Social Neuroscience Perspective on Physical Activity." *Journal of Sport & Exercise Psychology.* 30: 432–449.

Hamer, Dean H., Stella Hu, Victoria L. Magnuson, Nan Hu, and Angela M. L. Pattatucci. 1993. "A Linkage between DNA Markers on the X Chromosome and Male Sexual Orientation." *Science.* 261: 321–327.

Hamilton, William D. 1964. "Genetical Evolution of Social Behavior." *Journal of Theoretical Biology.* 7: 1–52.

Harris, Judith Rich. 1995. "Where Is the Child's Environment?: A Group Socialization Theory of Development." *Psychological Review.* 102: 458–489.

Harris, Judith Rich. 1998. *The Nurture Assumption: Why Children Turn Out the Way They Do.* New York: Free Press.

Harvey, P. H. and P. M. Bennett. 1985. "Sexual Dimorphism and Reproductive Strategies." Pp. 43–59 in Human Sexual Dimorphism, edited by J. Ghesquiere, R. D. Martin and F. Newcombe. London: Taylor & Francis.

Haselton, Martie G. and Daniel Nettle. 2006. "The Paranoid Optimist: An Integrative Evolutionary Model of Cognitive Biases." *Personality and Social Psychology Review.* 10: 47–66.

Henss, Ronald. 2000. "Waist-to-Hip Ratio and Female Attractiveness: Evidence from Photographic Stimuli and Methodological Considerations." *Personality and Individual Differences*. 28: 501–513.

Herrnstein, Richard J. and Charles Murray. 1994. *The Bell Curve: Intelligence and Class Structure in American Life*. New York: Free Press.

Hill, Kim and A. Magdalena Hurtado. 1996. *Ache Life History: The Ecology and Demography of a Foraging People*. New York: Aldine.

Hirschi, Travis and Michael J. Hindelang. 1977. "Intelligence and Delinquency: A Revisionist Review." *American Sociological Review*. 42: 571–587.

Holmstedt, Bo and Arne Fredga. 1981. "Sundry Episodes in the History of Coca and Cocaine." *Journal of Ethnopharmacology*. 3: 113–147.

Huang, Min-Hsiung and Robert M. Hauser. 1998. "Trends in Black-White Test-Score Differentials: II. The WORDSUM Vocabulary Test." Pp. 303–332 in *The Rising Curve: Long-Term Gains in IQ and Related Measure*, edited by Ulric Neisser. Washington, D.C.: American Psychological Association.

Hume, David. 1739. *A Treatise of Human Nature: Being an Attempt to Introduce the Experimental Method of Reasoning into Moral Subjects and Dialogues Concerning Natural Religion*. London: John Noon.

Hur, Yoon-Mi. 2007. "Stability of Genetic Influence on Morningness-Eveningness: A Cross-Sectional Examination of South Korean Twins from Preadolescence to Young Adulthood." *Journal of Sleep Research*. 16: 17–23.

Jackendoff, Ray. 2000. *Foundations of Language: Brain, Meaning, Grammar, Evolution*. Oxford: Oxford University Press.

Jensen, Arthur R. 1980. *Bias in Mental Testing*. New York: Free Press.

Jensen, Arthur R. 1998. *The* g *Factor: The Science of Mental Ability*. Westport: Praeger.

Jensen, Arthur R. and S. N. Sinha. 1993. "Physical Correlates of Human Intelligence." Pp. 139–242 in *Biological Approaches to the Study of Human Intelligence*, edited by Philip A. Vernon. Norwood: Ablex.

Jockin, Victor, Matt McGue, and David T. Lykken. 1996. "Personality and Divorce: A Genetic Analysis." *Journal of Personality and Social Psychology*. 71: 288–299.

Johnson, Wendy, Brian M. Hicks, Matt McGue, and William G. Iacono. 2009. "How Intelligence and Education Contribute to Substance

Use: Hints from the Minnesota Twin Family Study." *Intelligence.* 37: 613–624.

Kalmus, H. and D. B. Fry. 1980. "On Tune Deafness (Dysmelodia): Frequency, Development, Genetics and Musical Background." *Annals of Human Genetics.* 43: 369–382.

Kanazawa, Satoshi. 1998. "A Possible Solution to the Paradox of Voter Turnout." *Journal of Politics.* 60: 974–995.

Kanazawa, Satoshi. 2000. "A New Solution to the Collective Action Problem: The Paradox of Voter Turnout." *American Sociological Review.* 65: 433–442.

Kanazawa, Satoshi. 2001. "A Bit of Logic Goes a Long Way: A Reply to Sanderson." *Social Forces.* 80: 337–341.

Kanazawa, Satoshi. 2002. "Bowling with Our Imaginary Friends." *Evolution and Human Behavior.* 23: 167–171.

Kanazawa, Satoshi. 2004a. "The Savanna Principle." *Managerial and Decision Economics.* 25: 41–54.

Kanazawa, Satoshi. 2004b. "General Intelligence as a Domain-Specific Adaptation." *Psychological Review.* 111: 512–523.

Kanazawa, Satoshi. 2004c. "Social Sciences Are Branches of Biology." *Socio-Economic Review.* 2: 371–390.

Kanazawa, Satoshi. 2005. "An Empirical Test of a Possible Solution to 'the Central Theoretical Problem of Human Sociobiology.'" *Journal of Cultural and Evolutionary Psychology.* 3: 255–266.

Kanazawa, Satoshi. 2006a. "If the Truth Offends, It's Our Job to Offend." *Times Higher Education Supplement.* 15 December. 1773: 14.

Kanazawa, Satoshi. 2006b. "Why the Less Intelligent May Enjoy Television More than the More Intelligent." *Journal of Cultural and Evolutionary Psychology.* 4: 27–36.

Kanazawa, Satoshi. 2006c. "Mind the Gap . . . in Intelligence: Reexamining the Relationship between Inequality and Health." *British Journal of Health Psychology.* 11: 623–642.

Kanazawa, Satoshi. 2006d. "First, Kill All the Economists . . .": The Insufficiency of Microeconomics and the Need for Evolutionary Psychology in the Study of Management." *Managerial and Decision Economics.* 27: 95–101.

Kanazawa, Satoshi. 2006e. "IQ and the Wealth of States." *Intelligence.* 34: 593–600.

Kanazawa, Satoshi. 2009. "IQ and the Values of Nations." *Journal of Biosocial Science.* 41: 537–556.

Kanazawa, Satoshi. 2010a. "Why Liberals and Atheists Are More Intelligent." *Social Psychology Quarterly.* 73: 33–57.

Kanazawa, Satoshi. 2010b. "Evolutionary Psychology and Intelligence Research." *American Psychologist.* 65: 279–289.

Kanazawa, Satoshi. 2011. "Intelligence and Physical Attractiveness." *Intelligence.* 39: 7–14.

Kanazawa, Satoshi. Forthcoming. "Intelligence and Homosexuality." *Journal of Biosocial Science.*

Kanazawa, Satoshi and Jody L. Kovar. 2004. "Why Beautiful People Are More Intelligent." *Intelligence.* 32: 227–243.

Kanazawa, Satoshi and Deanna L. Novak. 2005. "Human Sexual Dimorphism in Size May Be Triggered by Environmental Cues." *Journal of Biosocial Science* 37: 657–665.

Kanazawa, Satoshi and Kaja Perina. 2009. "Why Night Owls Are More Intelligent." *Personality and Individual Differences.* 47: 685–690.

Kanazawa, Satoshi and Kaja Perina. Forthcoming. "Why More Intelligent Individuals Like Classical Music." *Journal of Behavior Decision Making.*

Kanazawa, Satoshi and Diane J. Reyniers. 2009. "The Role of Height in the Sex Difference in Intelligence." *American Journal of Psychology.* 122: 524–536.

Kanazawa, Satoshi and Mary C. Still. 1999. "Why Monogamy?" *Social Forces.* 78: 25–50.

Kaneshiro, Bliss, Jeffrey T. Jensen, Nichole E. Carlson, S. Marie Harvey, Mark D. Nichols, and Alison B. Edelman. 2008. "Body Mass Index and Sexual Behavior." *Obstetrics & Gynecology.* 112: 586–592.

Kenrick, Douglas T., Sara E. Gutierres, and Laurie L. Goldberg. 1989. "Influence of Popular Erotica on Judgments of Strangers and Mates." *Journal of Experimental Social Psychology.* 25: 159–167.

King, David P. and Joseph S. Takahashi. 2000. "Molecular Genetics of Circadian Rhythms in Mammals." *Annual Review of Neuroscience.* 23: 713–742.

Kingma, E. M., L. M. Tak, M. Huisman, and J.G.M. Rosmalen. 2009. "Intelligence Is Negatively Associated with the Number of Functional Somatic Symptoms." *Journal of Epidemiology and Community Health.* 63: 900–906.

Kirk, K. M., J. M. Bailey, M. P. Dunne, and N. G. Martin. 2000. "Measurement Models for Sexual Orientation in a Community of Twin Sample." *Behavior Genetics.* 30: 345–356.

Kirkpatrick, Lee A. 2005. *Attachment, Evolution, and the Psychology of Religion.* New York: Guilford.

Khan, Saad M. and Sumanta N. Pattanaik. 2004. "Modelling Blue Shift in Moonlit Scenes Using Rod Cone Interaction." *Journal of Vision.* 4 (8): 316.

Klein, David C., Robert Y. Moore, and Steven M. Reppert. 1991. *Suprachiasmatic Nucleus: The Mind's Clock.* New York: Oxford University Press.

Kluegel, James R. and Eliot R. Smith. 1986. *Beliefs about Inequality: Americans' View of What Is and What Ought to Be.* New York: Aldine.

Koenig, Laura B., Matt McGue, Robert F. Krueger, and Thomas J. Bouchard, Jr. 2005. "Genetic and Environmental Influences on Religiousness: Findings for Retrospective and Current Religiousness Ratings." *Journal of Personality.* 73: 471–488.

Kohler, Hans-Peter, Joseph L. Rodgers, and Kaare Christensen. 1999. "Is Fertility Behavior in Our Genes?: Findings from a Danish Twin Study." *Population and Development Review.* 25: 253–288.

Kohler, Hans-Peter, Joseph Lee Rodgers, Warren B. Miller, Axel Skytthe, and Kaare Christensen. 2006. "Bio-Social Determinants of Fertility." *International Journal of Andrology.* 29: 46–53.

Lake, Celinda C. and Vincent J. Breglio. 1992. "Different Voices, Different Views: The Politics of Gender." Pp. 178–201 in *The American Woman, 1992–93: A Status Report,* edited by Paula Ries and Anne J. Stone. New York: W. W. Norton.

Lee, Richard Borshay. 1979. *The !Kung San: Men, Women, and Work in a Foraging Society.* Cambridge: Cambridge University Press.

Lehrke, Robert. 1972. "A Theory of X-Linkage of Major Intellectual Traits." *American Journal of Mental Deficiency.* 76: 611–619.

Lehrke, Robert. 1997. *Sex Linkage of Intelligence: The X-Factor.* Westport, Conn.: Praeger.

Leighton, Donna Robbins. 1987. "Gibbons: Territoriality and Monogamy." Pp. 135–145 in *Primate Societies,* edited by Barbara Smuts, Dorothy L. Cheney, Robert M. Seyfarth, Richard Wrangham, and Thomas T. Struhsaker. Chicago: University of Chicago Press.

Leinonen, Lea, Ilkka Linnankoski, Maija-Liisa Laakso, and Reijo Aulanko. 1991. "Vocal Communication between Species: Man and Macaque." *Language & Communication.* 11: 241–262.

Leinonen, Lea, Maija-Liisa Laakso, Synnöve Carlson, and Ilkka Linnankoski. 2003. "Shared Means and Meanings in Vocal Expression of Man and Macaque." *Logopedics Phoniatrics Vocology.* 28: 53–61.

Lenski, Gerhard E. 1966. *Power and Privilege: A Theory of Social Stratification.* Chapel Hill: University of North Carolina Press.

Leutenegger, Walter and James T. Kelly. 1977. "Relationship of Sexual Dimorphism in Canine Size and Body Size to Social, Behavioral, and Ecological Correlates in Anthropoid Primates." *Primates.* 18: 117–136.

Levinson, David (editor in chief). 1991–1995. *Encyclopedia of World Cultures.* (10 volumes.) Boston: G. K. Hall.

Linnankoski, Ilkka., Maija Laakso, Reijo Aulanko, and Lea Leinonen. 1994. "Recognition of Emotions in Macaque Vocalizations by Children and Adults." *Language & Communication.* 14: 183–192.

Lubinski, David, Camilla P. Benbow, Rose Mary Webb, and April Bleske-Rechek. 2006. "Tracking Exceptional Human Capital over Two Decades." *Psychological Science.* 17: 194–199.

Lynam, Donald, Terrie E. Moffitt, and Magda Stouthamer-Loeber. 1993. "Explaining the Relation between IQ and Delinquency: Class, Race, Test Motivation, School Failure, or Self Control?" *Journal of Abnormal Psychology.* 102: 187–196.

Lynn, Michael and Barbara A. Shurgot. 1984. "Responses to Lonely Hearts Advertisements: Effects of Reported Physical Attractiveness, Physique, and Coloration." *Personality and Social Psychology Bulletin.* 10: 349–357.

Lynn, Richard. 1982. "IQ in Japan and the United States Shows a Growing Disparity." *Nature.* 297: 222–223.

Lynn, Richard. 1990. "The Role of Nutrition in the Secular Increases of Intelligence." *Personality and Individual Differences.* 11: 273–286.

Lynn, Richard. 1998. "In Support of the Nutrition Theory." Pp. 207–215 in *The Rising Curve: Long-Term Gains in IQ and Related Measures,* edited by Ulric Neisser. Washington, D.C.: American Psychological Association.

Lynn, Richard and John Harvey. 2008. "The Decline of the World's IQ." *Intelligence.* 36: 112–120.

Lynn, Richard, John Harvey, and Helmuth Nyborg. 2009. "Average Intelligence Predicts Atheism Rates Across 137 Nations." *Intelligence.* 37: 11–15.

Malamuth, Neil M. 1996. "Sexually Explicit Media, Gender Differences, and Evolutionary Theory." *Journal of Communication*. 46: 8–31.

McGrew, W. C. 1992. "Culture in Nonhuman Primates?" *Annual Review of Anthropology*. 27: 301–328.

McGue, Matt and David T. Lykken. 1992. "Genetic Influence on Risk of Divorce." *Psychological Science*. 3: 368–373.

Merrill, Maud A. 1938. "The Significance of IQ's on the Revised Stanford-Binet Scales." *Journal of Educational Psychology*. 29: 641–651.

Miller, Alan S. and John P. Hoffmann. 1995. "Risk and Religion: An Explanation of Gender Differences in Religiosity." *Journal for the Scientific Study of Religion*. 34: 63–75.

Miller, Alan S. and Satoshi Kanazawa. 2007. *Why Beautiful People Have More Daughters*. New York: Penguin.

Miller, Alan S. and Rodney Stark. 2002. "Gender and Religiousness: Can Socialization Explanations Be Saved?" *American Journal of Sociology*. 107: 1399–1423.

Miller, Geoffrey F. 2000a. *The Mating Mind: How Sexual Choice Shaped the Evolution of Human Nature*. New York: Doubleday.

Miller, Geoffrey. 2000b. "Sexual Selection for Indicators of Intelligence." Pp. 260–275 in *The Nature of Intelligence*, edited by Gregory R. Bock, Jamie A. Goode, and Kate Webb. New York: Wiley.

Miller, Geoffrey. 2009. *Spent: Sex, Evolution, and Consumer Behavior*. New York: Viking.

Miner, John B. 1957. *Intelligence in the United States: A Survey—with Conclusions for Manpower Utilization in Education and Employment*. New York: Springer.

Mithen, Steven. 2005. *The Singing Neanderthals: The Origins of Music, Language, Mind and Body*. London: Weidenfeld & Nicholson.

Moffitt, Terrie E. 1990. "The Neuropsychology of Delinquency: A Critical Review of Theory and Research." *Crime and Justice: An Annual Review of Research*. 12: 99–169.

Moffitt, Terrie E., William F. Gabrielli, Sarnoff A. Mednick, and Fini Schulsinger. 1981. "Socioeconomic Status, IQ, and Delinquency." *Journal of Abnormal Psychology*. 90: 152–156.

Moffitt, Terrie E. and Phil A. Silva. 1988. "IQ and Delinquency: A Direct Test of the Differential Detection Hypothesis." *Journal of Abnormal Psychology*. 97: 330–333.

Moore, George Edward. 1903. *Principia Ethica*. Cambridge: Cambridge University Press.

Murphy, Raegan. 2011. "The Lynn-Flynn Effect: How to Explain It?" *Personality and Individual Differences.*

Murray, Charles. 1997. *What It Means to Be a Libertarian: A Personal Interpretation.* New York: Broadway Books.

Murray, Charles. 2003. *Human Accomplishment: The Pursuit of Excellence in the Arts and Sciences, 800 B.C. to 1950.* New York: Perennial.

Mustanski, Brian S., Meredith L. Chivers, and J. Michael Bailey. 2002. "A Critical Review of Recent Biological Research on Human Sexual Orientation." *Annual Review of Sex Research.* 13: 89–140.

Nesse, Randolph M. 2001. "The Smoke Detector Principle: Natural Selection and the Regulation of Defensive Responses." *Annals of the New York Academy of Sciences.* 935: 75–85.

Nettl, Bruno. 1983. *The Study of Ethnomusicology: Twenty-nine Issues and Concepts.* Urbana: University of Illinois Press.

Nijenhuis, Jan te. 2011. "The Lynn-Flynn Effect in Korea." *Personality and Individual Differences.*

Nizynska, Joanna. 2010. "The Politics of Mourning and the Crisis of Poland's Symbolic Language after April 10." *East European Politics and Societies.* 24: 467–479.

Orians, Gordon H. and Judith H. Heerwagen. 1992. "Evolved Responses to Landscapes." Pp. 555–579 in *The Adapted Mind: Evolutionary Psychology and the Generation of Culture,* edited by Jerome H. Barkow, Leda Cosmides, and John Tooby. New York: Oxford University Press.

Pendergrast, M. (1999). *Uncommon Grounds: The History of Coffee and How It Transformed Our World.* New York: Basic Books.

Pérusse, Daniel. 1993. "Cultural and Reproductive Success in Industrial Societies: Testing the Relationship at the Proximate and Ultimate Levels." *Behavioral and Brain Sciences.* 16: 267–322.

Pickford, Martin. 1986. "On the Origins of Body Size Dimorphism in Primates." Pp. 77–91 in *Sexual Dimorphism in Living and Fossil Primates,* edited by Martin Pickford and Brunetto Chiarelli. Florence: Il Sedicesimo.

Pinker, Steven. 2002. *The Blank Slate: The Modern Denial of Human Nature.* London: Penguin.

Plotnik, Joshua, Frans B. M. de Waal, and Diana Reiss. 2006. "Self-Recognition in an Asian Elephant." *Proceedings of the National Academy of Sciences.* 103: 17053–17057.

Profant, Judi and Joel E. Dimsdale. 1999. "Race and Diurnal Blood Pressure Patterns." *Hypertension.* 33: 1099–1104.

Reiss, Diana and Lori Marino. 2001. "Mirror Self-Recognition in the Bottlenose Dolphin: A Case of Cognitive Convergence." *Proceedings of the National Academy of Sciences.* 98: 5937–5942.

Reiss, Diana and Brenda McCowan. 1993. "Spontaneous Vocal Mimicry and Production by Bottlenose Dolphins (*Tursiops truncatus*): Evidence for Vocal Learning." *Journal of Comparative Psychology.* 107: 301–312.

Rentfrow, Peter J. and Samuel D. Gosling. 2003. "The Do Re Mi's of Everyday Life: The Structure and Personality Correlates of Music Preference." *Journal of Personality and Social Psychology.* 84: 1236–1256.

Ridley, Matt. 1993. *The Red Queen: Sex and the Evolution of Human Nature.* New York: Penguin.

Ridley, Matt. 1996. *The Origins of Virtue: Human Instincts and the Evolution of Cooperation.* New York: Viking.

Ridley, Matt. 1999. *Genome: The Autobiography of a Species in 23 Chapters.* New York: Perennial.

Roberts, Richard D. and Patrick C. Kyllonen. 1999. "Morningness-Eveningness and Intelligence: Early to Bed, Early to Rise Will Likely Make You Anything but Wise!" *Personality and Individual Differences.* 27: 1123–1133.

Romero, Gorge A. and Aaron T. Goetz. 2010. "Sexually Insatiable Women: Evidence for the Savanna-IQ Interaction Hypothesis with Pornography." Department of Psychology. California State University.

Ross, Callum F. 2000. "Into the Light: The Origin of Anthropoidea." *Annual Review of Anthropology.* 29: 147–194.

Rowe, David C. 1994. *The Limits of Family Influence: Genes, Experience, and Behavior.* New York: Guilford.

Sally, David. 1995. "Conversation and Cooperation in Social Dilemmas: A Meta-analysis of Experiments from 1958 to 1992." *Rationality and Society.* 7: 58–92.

Savage-Rumbaugh, Sue and Roger Lewin. 1994. *Kanzi: The Ape at the Brink of the Human Mind.* New York: Wiley.

Seielstad, Mark T., Eric Minch, and L. Luca Cavalli-Sforza. 1998. "Genetic Evidence for a Higher Female Migration Rate in Humans." *Nature Genetics.* 20: 278–280.

Shapiro, Robert Y. and Harpreet Mahajan. 1986. "Gender Differences in Policy Preferences: A Summary of Trends from the 1960s to the 1980s." *Public Opinion Quarterly.* 50: 42–61.

Shaw, Bernard. 1957. *Man and Superman*. New York: Penguin.

Shayer, Michael and Denise Ginsburg. 2009. "Thirty Years On—A Large Anti-Flynn Effect? (II): 13- and 14-Year-Olds. Piagetian Tests of Formal Operations Norms 1976–2006/7." *British Journal of Educational Psychology.* 79: 409–418.

Shayer, Michael, Denise Ginsburg, and Robert Coe. 2007. "Thirty Years On—A Large Anti-Flynn Effect? The Piagetian Test *Volume & Heaviness* Norms 1975–2003." *British Journal of Educational Psychology.* 77: 25–41.

Shepard, Roger N. 1994. "Perceptual-Cognitive Universals as Reflections of the World." *Psychonomic Bulletin & Review.* 1: 2–28.

Shin, Jae Chul, Hirohisa Yaguchi, and Satoshi Shioiri. 2004. "Change of Color Appearance in Photopic, Mesopic and Scotopic Vision." *Optical Review.* 11: 265–271.

Silventoinen, Karri, Sampo Sammalisto, Markus Perola, Dorret I. Boomsma, Belinda K. Cornes, Chyana Davis, Leo Dunkel, Marlies de Lange, Jennifer R. Harris, Jacob V. B. Hjelmborg, Michelle Luciano, Nicholas G. Martin, Jakob Mortensen, Lorenza Nisticò, Nancy L. Pedersen, Axel Skytthe, Tim D. Spector, Maria Antonietta Stazi, Gonneke Willemsen, and Jaakko Kaprio. 2003. "Heritability of Adult Body Height: A Comparative Study of Twin Cohorts in Eight Countries." *Twin Research.* 6: 399–408.

Silverman, Irwin, Jean Choi, Angie Mackewn, Maryanne Fisher, Judy Moro, and Esther Olshansky. 2000. "Evolved Mechanisms Underlying Wayfinding: Further Studies on the Hunter-Gatherer Theory of Spatial Sex Differences." *Evolution and Human Behavior.* 21: 201–213.

Singh, Devendra. 1993. "Adaptive Significance of Waist-to-Hip Ratio and Female Physical Attractiveness." *Journal of Personality and Social Psychology.* 65: 293–307.

Singh, Devendra. 1994. "Is Thin Really Beautiful and Good? Relationship between Waist-to-Hip Ratio (WHR) and Female Attractiveness." *Personality and Individual Differences.* 16: 123–132.

Singh, D. and S. Luis. 1995. "Ethnic and Gender Consensus for the Effect of Waist-to-Hip Ratio on Judgments of Women's Attractiveness." *Human Nature.* 6: 51–65.

Singh, D. and R. K. Young. 1995. "Body Weight, Waist-to-Hip Ratio, Breasts, and Hips: Role in Judgments of Female Attractiveness and Desirability for Relationships." *Ethology and Sociobiology.* 16: 483–507.

Smith, E. O. 1999. "Evolution, Substance Abuse, and Addiction." Pp. 375–400 in *Evolutionary Medicine*, edited by Wenda R. Trevathan, E. O. Smith, and James J. McKenna. New York: Oxford University Press.

Smith, Robert L. 1984. "Human Sperm Competition." Pp. 601–659 in *Sperm Competition and the Evolution of Mating Systems*, edited by Robert L. Smith. New York: Academic Press.

Smuts, Barbara. 1985. *Sex and Friendship in Baboons.* New York: Aldine.

Spearman, C. (1904). "General Intelligence, Objectively Determined and Measured." *American Journal of Psychology*, 15, 201–293.

Stanovich, Keith E., Anne E. Cunningham and Dorothy J. Feeman. 1984. "Intelligence, Cognitive Skills, and Early Reading Progress." *Reading Research Quarterly.* 19: 278–303.

Stigler, George J. and Gary S. Becker. 1977. "De Gustibus Non Est Disputandum." *American Economic Review.* 67: 76–90.

Sundet, Jon Martin, Dag G. Barlaug, and Tore M. Torjussen. 2004. "The End of the Flynn Effect?: A Study of Secular Trends in Mean Intelligence Test Scores of Norwegian Conscripts during Half a Century." *Intelligence.* 32: 349–362.

Sundquist, James L. 1983. *Dynamics of the Party System*, Revised Edition. Washington, D.C.: Brookings Institution.

Symons, Donald. 1979. *The Evolution of Human Sexuality.* Oxford: Oxford University Press.

Symons, Donald. 1990. "Adaptiveness and Adaptation." *Ethology and Sociobiology.* 11: 427–444.

Teasdale, Thomas W. and David R. Owen. 2005. "A Long-Term Rise and Recent Decline in Intelligence Test Performance: The Flynn Effect in Reverse." *Personality and Individual Differences.* 39: 837–843.

Thornhill, Randy. 1980. "Rape in *Panorpa* Scorpionflies and a General Rape Hypothesis." *Animal Behaviour.* 28: 52–59.

Thornhill, Randy and Craig T. Palmer. 2000. *A Natural History of Rape: Biological Bases of Sexual Coercion.* Cambridge: MIT Press.

Thornhill, Randy and Nancy Wilmsen Thornhill. 1983. "Human Rape: An Evolutionary Analysis." *Ethology and Sociobiology.* 4: 137–173.

Tooby, John and Leda Cosmides. 1990. "The Past Explains the Present: Emotional Adaptations and the Structure of Ancestral Environments. *Ethology and Sociobiology*, 11, 375–424.

Tovée, Martin J. and Piers L. Cornelissen. 2001. "Female and Male Perceptions of Female Physical Attractiveness in Front-View and Profile." *British Journal of Psychology.* 92: 391–402.

Trivers, Robert L. 1972. "Parental Investment and Sexual Selection." Pp. 136–179 in *Sexual Selection and the Descent of Man 1871–1971*, edited by Bernard Campbell. Chicago: Aldine.

Turkheimer, Eric. 2000. "Three Laws of Behavior Genetics and What They Mean." *Current Directions in Psychological Science.* 9: 160–164.

Turner, Gillian. 1996a. "Finding Genes on the X Chromosome by Which *Homo* May Have Become *Sapiens.*" *American Journal of Human Genetics.* 58: 1109–1110.

Turner, Gillian. 1996b. "Intelligence and the X Chromosome." *Lancet.* 347: 1814–1815.

Vallee, Bert L. 1998. "Alcohol in the Western World." *Scientific American.* 278 (6): 80–85.

van Beest, Ilja and Kipling D. Williams. 2006. "When Inclusion Costs and Ostracism Pays, Ostracism Still Hurts." *Journal of Personality and Social Psychology.* 91: 918–928.

van den Berghe, Pierre L. 1990. "From the Popocatepetl to the Limpopo." Pp. 410–431 in *Authors of Their Own Lives: Intellectual Autobiographies by Twenty American Sociologists*, edited by Bennett M. Berger. Berkeley: University of California Press.

Vanhanen, Tatu. 2003. *Democratization: A Comparative Analysis of 170 Countries.* New York: Routledge.

van Lawick-Goodall, Jane. 1968. "Tool-Using Bird: The Egyptian Vulture." *National Geographic.* 133 630–641.

Vitaterna, Martha Hotz, Joseph S. Takahashi, and Fred W. Turek. 2001. "Overview of Circadian Rhythms." *Alcohol Research and Health.* 25: 85–93.

Volk, Tony and Jeremy Atkinson. 2008. "Is Child Death the Crucible of Human Evolution?" *Journal of Social, Evolutionary, and Cultural Psychology.* 2: 247–260.

de Waal, Frans B. M. 1982. *Chimpanzee Politics: Power and Sex among Apes.* London: Jonathan Cape.

de Waal, Frans B. M. 1989. "Food Sharing and Reciprocal Obligations among Chimpanzees." *Journal of Human Evolution.* 18: 433–459.

de Waal, Frans B. M. 1992. "Appeasement, Celebration, and Food Sharing in the Two *Pan* Species." Pp. 37–50 in *Topics in primatology: Human origins*, edited by Toshisada Nishida, W. C. McGrew, and Peter Marler. Tokyo: University of Tokyo Press.

de Waal, Frans B. M. 1995. "Bonobo Sex and Society." *Scientific American.* 272 (3): 82–88.

de Waal, Frans. 1996. *Good Natured: The Origins of Right and Wrong in Humans and Other Animals.* Cambridge: Harvard University Press.

de Waal, Frans B. M., Lesleigh M. Luttrell, and M. Eloise Canfield. 1993. "Preliminary Data on Voluntary Food Sharing in Brown Capuchin Monkeys." *American Journal of Primatology.* 29: 73–78.

Weiss, Alexander, Alicia Morales, and W. Jake Jacobs. 2003. "Place Learning in Virtual Space IV: Spatial Navigation and General Intelligence Appear Independent." Department of Psychology. University of Arizona.

Whitmeyer, Joseph M. 1997. "Endogamy as a Basis for Ethnic Behavior." *Sociological Theory.* 15: 162–178.

Whitten, Norman E. Jr. 1976. *Sacha Runa: Ethnicity and Adaptation of Ecuadorian Jungle Quichua.* Urbana: University of Illinois Press.

Wilbert, Johannes. 1991. "Does Pharmacology Corroborate the Nicotine Therapy and Practices of South American Shamanism?" *Journal of Ethnopharmacology.* 32: 179–186.

Wilson, Edward O. 1975. *Sociobiology: The New Synthesis.* Cambridge: Harvard University Press.

Wilson, Glenn and Qazi Rahman. 2005. *Born Gay: The Psychobiology of Sex Orientation.* London: Peter Owen.

Wilson, James Q. and Richard J. Herrnstein. 1985. *Crime and Human Nature: The Definitive Study of the Causes of Crime.* New York: Touchstone.

Wirls, Daniel. 1986. "Reinterpreting the Gender Gap." *Public Opinion Quarterly.* 50: 316–330.

Wolfgang, Marvin E., Robert M. Figlio, and Thorsten Sellin. 1972. *Delinquency in a Birth Cohort.* Chicago: University of Chicago Press.

Wolfle, Lee M. 1980. "The Enduring Effects of Education on Verbal Skills." *Sociology of Education.* 53: 104–114.

Wrangham, Richard W., W. C. McGrew, Frans B. M. de Waal, and Paul G. Heltne. 1994. *Chimpanzee Cultures.* Cambridge: Harvard University Press.

Wrangham, Richard and Dale Peterson. 1996. *Demonic Males: Apes and the Origins of Human Violence.* Boston: Mariner.

Wray, Alison. 1998. "Protolanguage as a Holistic System of Social Interaction." *Language and Communication*. 18: 47–67.

Wray, Alison. 2006. "Joining the Dots: The Evolutionary Picture of Language and Music." *Cambridge Archaeological Journal*. 16: 103–105.

Wright, Robert. 1994. *The Moral Animal: The New Science of Evolutionary Psychology.* New York: Vintage.

Yeudall, Lorne T., Delee Fromm-Auch, and Priscilla Davies. 1982. "Neuropsychological Impairment of Persistent Delinquency." *Journal of Nervous and Mental Diseases.* 170: 257–265.

Zagar, Robert and John D. Mead. 1983. "Analysis of Short Test Battery for Children." *Journal of Clinical Psychology.* 39: 590–597.

Zahavi, Amotz. 1975. "Mate Selection—Selection for a Handicap." *Journal of Theoretical Biology.* 53: 205–214.

Zahavi, Amotz and Avishag Zahavi. 1997. *The Handicap Principle: A Missing Piece of Darwin's Puzzle.* New York: Oxford University Press.

Zuberbühler, Klaus. 2002. "A Syntactic Rule in Forest Monkey Communication." *Animal Behaviour.* 63: 293–299.

Zuberbühler, Klaus. 2003. "Natural Semanticity in Wild Primates." Pp. 362–367 in *Animal Social Complexity: Intelligence, Culture, and Individualized Societies.* Edited by Frans B. M de Waal and Peter L. Tyack. Cambridge: Harvard University Press.

Index

Note: Page numbers followed by *f* indicate figures.